Pelican Books
**Weapons and Tactics**

Tom Wintringham was a leading left-wing journalist in the
1930s and a writer for the Left Book Club. He fought in
the Spanish Civil War and was killed early in the Second
World War.

John Blashford-Snell is the son of a Church of England
clergyman. Educated at Victoria College, Jersey, he was
commissioned into the Royal Engineers in 1957. His active
service took him all over the world, and in 1963 he
became an instructor at The Royal Military Academy,
Sandhurst. In this position he has organized over sixty
expeditions, taking part in a number of them. These have
included underwater archaeological quests, the first
descent and exploration of the Blue Nile, and the most
recent in 1969 when the Ethiopian government invited him
to explore an archipelago in the Red Sea.
 He was awarded the M.B.E. in 1969, and the Cuthbert
Peek Award by the Royal Geographical Society in 1970.
Major Blashford-Snell is married with two daughters.

*The authors and the publishers wish to point out that the
text of this book was set before October 1973. Revisions
in the light of events in the Middle East have proved
impossible to make without unduly delaying publication.*

# Weapons and Tactics

Tom Wintringham and
J. N. Blashford-Snell

Penguin Books

Penguin Books Ltd, Harmondsworth,
Middlesex, England
Penguin Books Inc., 7110 Ambassador Road,
Baltimore, Maryland 21207, U.S.A.
Penguin Books Australia Ltd, Ringwood,
Victoria, Australia

Part One first published by Faber and Faber 1943
Part Two first published 1973 in this edition
Part One copyright © Tom Wintringham, 1943
Part Two copyright © J. N. Blashford-Snell, 1973

Made and printed in Great Britain by
Richard Clay (The Chaucer Press) Ltd,
Bungay, Suffolk
Set in Linotype Times

More than most professions the military is forced to depend on intelligent interpretation of the past for signposts charting the future. Devoid of opportunity, in peace, for self-instruction through actual practice in his profession, the soldier makes maximum use of historical record in assuring the readiness of himself and his command to function efficiently in emergency. The facts derived from historical analysis he applies to conditions of the present and the proximate future, thus developing a synthesis of appropriate method, organization, and doctrine.

*– from the report presented to the Secretary of War at Washington on 30 June 1935, by*

GENERAL DOUGLAS MACARTHUR

# Contents

# Introduction

Two thousand years ago an armoured force was marching on foot towards what is now Sedan; the legions of the first Caesar were moving to the conquest of the Belgae. One thousand years ago an armoured force rode on horses towards the same place, from another direction; some knights who may have owed allegiance to the first King of the Germans, Otto the Great, were settling a little matter arising from the break-up of Charlemagne's almost united Europe. Two years ago an armoured force, speeding in vehicles through Sedan, began the conquest of most of modern Europe. This is one of the patterns that shape the history of war.

This history of weapons and tactics has been nearly three years in the writing. It was begun when German armoured divisions were storming towards Warsaw – the city from which, about three hundred years earlier, the last heavily armoured chivalry of Europe rode under Sobieski to repel the Turks. It is a book written while the clanging of armour is in all men's ears. It goes through thousands of years of the past to shape our understanding of the Panzer divisions of the present. It starts from the Siege of Troy in order to make clear the military meaning of the siege of Stalingrad.

This book has only one aim: that we should understand warfare and therefore win this war. It is not a complete history of weapons and tactics. It is a history of those changes of the past in weapons and tactics that seem to have a bearing on the present, on the revolution in warfare that has taken place in the last twenty-six years, since the first tank ploughed through the mud of the Somme – heading, by a freak of history, in the general direction of Sedan.

Our need to understand warfare is very urgent today. We, the British, as I write in the late summer of 1942, are losing our share of the war. We expected to take the offensive in 1942 – as in 1916 after less than two years of fighting we were able to exert enough clumsy force, on the Somme, to rock our enemies back on to the defensive along the whole Western Front. We expected that the mass of armoured machines we made, in the two years after Dunkirk, would prove to be of higher quality that the enemy's machines. We have been disappointed. Somewhere wrong ideas of how to fight have crippled us; somehow against almost unarmoured Japanese, as well as against heavily armoured Germans, our forces have not yet shown an efficiency equal to their courage. There are many things to be altered. If we are to do the alterations well we must plan them; that means we must fit them into a general idea of how to fight now and in the future; and that again means we should have a general idea of fighting in the past. If I dared, I would say we must have a theory – the word 'theory' is so much disliked by so many Englishmen, and is considered by them so 'unpractical', that I avoid it all I can; although I cannot see, myself, that it is very 'practical' to do things without knowing the theory of how to do them.

Our difficulty is not mainly in the design, manufacture and use of weapons as such. Most of our weapons are more nearly perfect, taken as 'things in themselves', than our enemies'. The difficulty lies in the systems and methods of their use in battle, in tactics, and in the design of weapons particularly suitable for use in the tactics now profitable.

Most modern weapons of which I have personal experience seem to me extremely easy to handle after a fashion, and not really hard to handle fairly well. A bullet, I believe, is usually easier to aim at a small target than a football is at a large one. It is certainly easier to aim than an arrow. The business of making a good black pattern on white paper is easier with a machine-gun than with a printing press. As for the mortars and anti-tank guns that modern armies use, they seem far more fool-proof than a vacuum cleaner and much less dangerous (to the man using them) than a motor-cycle. I have personally instructed the machine-gunners of three battalions of the International Brigades in Spain in the use of a new type of machine-gun, that neither they nor I had ever seen before, in a few hours of the day before those battalions went over the top; in those few hours the men seemed to get a rough but sufficient idea of the operation of these guns; they were able to work them almost as well as the other types, somewhat worn, to which they were more accustomed. Few soldiers, I think, find great difficulty in weapon-training, in learning to use the weapons of today. Even a tank, though tiring, is not very hard to drive. What is hard is to learn where and when to drive tanks, where, when and in what direction to site and to fire weapons.

It is even more difficult to learn how to combine the effect of a number of weapons, of different types, together with the movement of men who are using or serving these weapons, and of the machines that carry or pull them. The effective combination of fire and movement – soldiers will find it in their training manuals – is the essential problem of tactics. Weapons have no means apart from the use of weapons; separated from tactics they become heavy and knobbly things for tired men to carry or drag. It is impossible to learn the right use of weapons without at the same time learning tactics.

Soldiers learn tactics, how to interweave their fire and how to move under enemy fire, by actual experience in training fields and in battle. But they can also learn from reading about the past. No man knows his own trade or profession well unless he knows how people used to do things, not very different in essence from what he is trying to do, under the very different circumstances of past ages. When you are trying to learn to write, you are taught at school how men in the past have written. When you are trying to learn to fight, it is worth while learning something about the tools and machines with which men fought long ago, and their changing ways of using these things.

In any writing of history the writer must choose and arrange the facts described, according to the importance he gives them. I have done this, deliberately, to form an argument. By means of this argument I hope to clear a way towards a theory or doctrine of warfare that is in advance of the Nazi doctrine – closer to reality, more capable of victory.

In other words I do not hope merely to help the soldier and the junior officer to understand more easily the use of modern weapons, the tactics now taught to him, because he understands those of the past; I also hope that my summary of the changes occurring in the past will be of service, in Britain and America, to those who see that victory depends on hard thinking about the art and science of war, as well as on courage and devotion, obedience and initiative.

Some of my general argument about the development of war comes in the first chapter of this book. Its application to modern conditions comes in the last.* Some practical friends of mine, including some good soldiers, will ask themselves – or even ask me – why they should wade through the historical chapters between these. My answer is that no one can fully grasp a science or an art unless he gets a clear idea about its history.

Treat fighting simply as a job of work, if you like; it is skilled work and needs learning, like all the other jobs from A to Z, from advertising to zoo-keeping. I am sure I could find a complete history of advertising or aeronautics or algebra in any library. But war? A great historian, Oman, wrote a history of the art of war in the Middle Ages. A German soldier, Clausewitz, included a good deal of the history of war, up to Napoleon's time, in his study of it. We need far more than this. We need – do I not know it! – better history than I or anyone else can write when war presses essential and immediate tasks on all of us. But here is a beginning.

This book is not written for soldiers only, or mainly. Civilians can neither understand what is happening nor help, politically and militarily, towards victory unless they also have a general idea of war. And war today is a matter for every

*Chapter 10 in this Pelican edition.

man's concern. It seizes on all of us; it changes our work, our food, often our place of living, always our way of living. It darkens and destroys our cities; it separates us, in many cases, from our partners in marriage and from our children. A war such as now engages us has, for almost all of us, far more effect and importance than had most wars of the past, wars that were fought far away and scarcely rippled the surface of everyday life for those at home.

There still remain, in Britain and America, some who think of war as it used to be, a concern somewhat separate from the life of the nation, a business mainly for those who have chosen the profession of arms, something that cannot be understood by the mass of the people. War seems to these people a matter for the experts. Civilians, and the politicians who represent or misrepresent them, should interfere as little as possible with those engaged in the business of fighting. The untrained man, they think, cannot possibly understand even in outline the subtle art of warfare, this mystery to the learning of which experts have to give their best years, their whole lives.

There was a time in England when many of our trades and professions were organized in closed corporations, some of which were called guilds. A man entering a profession had to go through a period of apprenticeship that might last for many years. During this period he was supposed to be learning 'the mystery' of printing or of shoemaking. Our forefathers even made a mystery out of the getting and selling of fish, and one of the most handsome (and doubtless most beneficent) of the greater corporations of the City of London is the Honourable (or Worshipful – the full title escapes me) Company of Fishmongers. The army, in Britain, still retains some of this separate character of the medieval guild. It still wears unusual clothes on ceremonial occasions, or eats its meals with a ritual of which the origin is half forgotten, and its reasons for doing so are much the same as the reasons why the chairman (or Right Whale, or whatever his title is) of the Worshipful Fishmongers will wear on important occasions, at banquets of ceremony, golden chains – chains that are not, as might be imagined, symbols of the hard bondage of the sea, but are more probably symbols of the linked dependence of one man upon another that used to be embodied in the guilds.

To choose a more recent comparison, the British army often seems very like a collection of those trade unions that flourish mainly in the older trades, in which apprentices becoming 'worthy brothers' go through a complicated rigmarole of customary initiation; as part of this they pledge themselves not to reveal to the uninitiated the inherited secrets of their craft. Sometimes it seems as if, in our army, the skill and weapons of artillerymen or tankmen are carefully not revealed to the ordinary men of the infantry, the uninitiated – particularly carefully not revealed to recruits. And never to civilians.

There was a time when all sciences were secret or half-secret things. Members of the Royal Society, nearly 300 years ago, peering down the clumsy microscope invented by a fantastic Dutchman (who claimed to see in a drop of water strange tiny creatures not mentioned in the Bible or in Aristotle, and therefore somehow indecent) kept their discussions mainly within their own charmed circle; the vulgar and unlettered could not understand or even be interested in the mysteries that constituted science.

Many things have changed since those days. There are still trade secrets, there are still craft skills. Science still finds its way to ideas that are hard to understand. But any schoolboy with a mechanical turn of mind can read weekly, in journals specially written for him, of all the technique of making and getting and shifting things; an unforgettable film shows us the work of drifters and trawlers, work that was once part of the secret art of a Worshipful Company. And when an Einstein compresses space and time, magnetism and motion into some new and very difficult formula of relativity, newspapers that boast of two million readers try to explain his theory – mainly in sentences not more than seven words long.

But boys' papers that describe with the most beautiful lucidity the workings of a power station or of a coal pit, when they write of war go in for romantic heroism, of a sort that became slightly old-fashioned in the fourteenth century, or else bewilder the ardent young with nonsense about death rays (and worse nonsense about bomb-sights that make it certain that bombs will hit their targets). As for what the 'adult' press does when it tries to describe and analyse warfare – I had better not give my view on that.

The view that war is a mystery, a secret business, does not suit a democracy. It is absurd that men should be encouraged to educate themselves in ways of getting a living and discouraged from finding out about ways in which they may get their deaths. And when war becomes the principal business of men and women in a population of many millions, they need to have at least a general idea of this business; otherwise their best energies and most vigorous enthusiasms may go adrift in uselessness.

So much of the energy of men and nations has been devoted to warfare in the past three thousand years that even the shortest description of what has happened would fill many books of this size. The thing attempted here is smaller; to describe the main lines of development of the weapons with which men have fought – on land – during most of these years, and some of the reasons why these weapons have changed, and why there have occurred changes in the methods of their use.

This leaves out of account a great many things that have been important in warfare, and are still important. Leadership is important; luck is important; sea-power and sea-fighting are important; and there have been in the past and may again be today beliefs, faiths, and desires, religious and political, that have moved nations to decisive effort in warfare. These things do not come into our picture, in this book.

Nor is there any attempt here to make a complete list of every weapon that has been or is being used in war. This is not a catalogue. Too many of the books that describe weapons of the past consist of a simple catalogue hung round with a few stray anecdotes. This is an attempt to explain, not enumerate.

The omission of naval warfare and naval weapons is not due

to a disregard for their importance. It is due to the length of this book and the author's inexperience of war at sea.

I write, then, for civilians and soldiers, being sure that in a democracy at war the part of it that does the fighting and the part that works at home need to have a unity of understanding. They must work together as a team; the civilians must support the services as backs and half-backs support the forward line, the goal-scorers, in football. They cannot have this understanding without knowledge of past wars. The stevedore deciding how to dump part of a cargo, and the Prime Minister deciding on the aims to be set before whole armies, both need to have a general idea of the sort of war they are waging, the 'priorities' involved, the first things we need to have and to do. So does the sergeant left alone to 'carry on', his officer wounded or separated from him in the swaying flux of modern battle.

I am not saying that the reading of history and the discussion of theory is the only, or the best, way to learn war. Fighting is the best way. I write before we have invaded Europe, but I hope we shall do so – for many reasons, including the hope that thereby we learn how to fight better than the Nazis, and apply this knowledge to the training and retraining of forces capable of final victory.

Yet whatever fighting lies ahead, and whatever training, will be done better if our practical soldiers see the need for theory. That is why I labour this point. I am English and therefore practical myself. I prefer training or commanding troops to reading up history. In Spain I helped to train several battalions of the International Brigades and a number of officers for those Brigades. More recently at the Osterley School for the Home Guard I helped to train 5,000 officers and men; at the War Office School, when Osterley was officially taken over, several thousand more. But in this rewarding practical work I found one great lack: the lack of a theory of warfare.

How can you train troops, or command them, if you have no theoretical view of war and its development? Obviously you *can* do it; it is being done. Nothing else but exactly that is being done in most of the British army. But how do you do it? To judge from my own experience, you do it by having in your mind a lot of ideas about war that are jumbled up together, without order or logic. You do some of the things that everyone else does; you follow the manuals: or you have 'pet ideas' of your own; you 'do your best'. You have some personal experience of past fighting, that colours and perhaps over-colours the picture of war you have in mind. You rely on your imagination – good servant, bad master, is friend imagination. And you try to decide by the light of reason between different ideas or 'tips' that come your way.

It is not hard for anyone to see that if you go about the job in this way, the resulting mixture will have a great deal of the past in it. Your own experience, perhaps, of past wars or past campaigns; the manuals' summaries of others' past experiences; the traditional 'we have always done it this way' – these things are likely to dominate what you do. Should they?

I say they should not – unless they are 'rationalized', straightened out by reason. I go driving back into the past, in this book, in order to get the past into proper order, in my own mind and in others, and in order to use it for a proper purpose. That purpose is to learn the ways of change. It was the elder (and greater) von Moltke who used to repeat: 'In war, only change is admissible.'

History ought to tell us how change happens, what are the likely developments today and tomorrow, what 'patterns' are likely to be repeated – like the pattern of armoured soldiers marching, then riding, then driving vehicles towards Sedan, repeated at intervals of a thousand years – or can be made to repeat themselves with changed colours or in some new shape. If this history of mine brings us in any way nearer that understanding, it has been worth doing.

T.W.

*London, August 1942*

**Part One by Tom Wintringham**

# 1. The Theory of the Thing

Battle of Hastings, 1066, is the date that everybody in England knows. It does not come at the beginning of the history of weapons and tactics – any more than it comes at the beginning of the history of England – but we can start this chapter with it because it provides us with an example of the sort of thing to be explained, if we are to understand warfare. In that year Duke William of Normandy defeated and killed King Harold of England, and achieved the Conquest. Why did it happen? In school-book histories, children get the impression that the Norman leader was a better soldier than the English; that the Battle of Hastings was won by his leadership, and by the cleverness and the fierce courage of the Normans. All this is, within its limits, true. But something else is left out; and often it is left out of the better histories that are not written simply for school children; it is for example left out of one of the best of these histories, John Richard Green's *The Conquest of England*. A new way of fighting, of handling weapons, of protecting the men who handle weapons – new in England, although it had been developing for hundreds of years in other parts of the world – met and defeated an old way of fighting.

Quite true that William was a better soldier than Harold, though not a braver one. True that the Norman army man-oeuvred better than the English, but this was less due to their inborn or acquired cleverness than to the fact that the Norman army had as its shock troops an armoured cavalry; the English fought on foot.

There were reasons for this difference. The two armies' ways of fighting derived, if you go back far enough, from their ways of living, from systems of society or a lack of system or a mixture of systems. And these in turn had embedded in them all the peaceful or warlike history of the peoples that formed the armies. But this connection between ways of living and ways of fighting (between economics and politics on the one hand and warfare on the other) is not the theme of this book. What matters to us is the fact that war was changing, as far back as 1066; a man on a horse could wear heavier armour (and carry heavier weapons) than a man on foot, and therefore the man on horseback – chivalry – conquered England.

Armour helped to conquer England then. If Britain is invaded in the present war, it will be largely with armour that the invaders try to conquer us. The heavy cavalry of the Norman Conquest were the Panzer divisions of their day. They had more mobility, tactically, as well as more hitting-power than the English infantry they beat. And they had more protection.

Mobility, hitting-power, protection – these are, and always have been, some of the keys to victory.* Where other things – surprise, or concentration at the decisive place and time, or skilful manoeuvre – have won battles, these things have usually derived directly from superiority in mobility, hitting-power and protection, or from superiority in one or two of these qualities.

It would be foolish to ask which is the more important of these three qualities. There have been phases of war in which one was all-important, and the other two did not matter much. But usually these three qualities have been interlinked, in 'new' armies that have altered accepted ways of fighting. That makes the past of warfare difficult to disentangle; it would be far easier if there were only one outstanding factor or quality that always mattered more than others.

Then we could analyse the whole past of war according to

*These three are, I believe, accepted by British military opinion today, together with morale, as the most important factors in battle. With morale – with which I do not deal here, because this is a book on weapons and tactics – they are stated as such in the best expression of British military thought recently made available, the Lees Knowles Lectures for 1942 by Major-General G. M. Lindsay.

the development of this single quality. If mobility were the only thing that mattered, for example, we could split up warfare according to the speed of movement of troops. We should first have footslogging war, then horse-riding war, then railway war, then petrol war. And within these divisions or sections of warfare we could trace developments; the training that made the Roman Legions able to outmarch their opponents, the horsemanship of the Saracens or Mongols, the breakdown of the countries with inadequate railways in the 'railway war' of 1914–18 (Serbia, Rumania, Russia, Turkey, Bulgaria, Austria), and today the development of the petrol engine to dominate war. But in fact mobility is not the one thing that matters, always, in all war. And the analysis I have sketched breaks down when you look at the history of war on foot and war on horses.

There has not been a plain and simple progress from marching to riding. Hitting-power and protection have complicated this; there have been successive 'cavalry periods', in each of which cavalry has played a rather different part in the game. There have been other periods when, for reasons connected with hitting-power and protection, cavalry has mattered very little.

An attempt has been made to analyse the past of war mainly according to the hitting-power of armies, the quality that comes second in my list of three 'keys to victory' (though it might equally well come first). Major-General J. F. C. Fuller has tried it, splitting war into alternating 'shock cycles' and 'projectile cycles'. The reader who wishes to follow this idea from the year 1100 B.C. into the future – to A.D. 2050 – will find all these centuries compressed into three pages of General Fuller's book, *The Dragon's Teeth* (pp. 227–9).

I believe that his division of the periods of warfare (into those in which men fight mainly at close quarters with 'shock weapons' and those in which they fight mainly at long range with missiles and projectiles) is of great value; but it is incomplete and misleading as a guiding thread to the study of all the history of war. This guiding thread ought to show us a pattern that links mobility and protection, as well as fire-power, into a comprehensible shape for the various periods of war. I believe we find it most easily when we start with protection, and link with this quality the other two.

Perhaps this seems to me the essential background pattern of war because I am English, and have been living and working on this book in an England threatened by invasion. It is natural for me to think of past invasions. England has not often been invaded successfully from the Continent. The last time it was done, by Dutch William of Orange in the 'Glorious Revolution' of 1688, most of the people of the country welcomed the 'invaders'; it scarcely ranks as a foreign conquest. The foreign conquests that matter in our history are those of Julius Caesar, and of Duke William of Normandy.

Caesar led an armoured force to these islands; William led an armoured force; in 1940 and 1941 it was armoured force that threatened us. At intervals of a thousand years or so, these armoured forces, so different and yet in some ways so similar, seem to me wave-crests in the history of war. There are other wave-crests; but these are the ones that press on my mind in England today.

A primary connection between mobility and protection comes into the pattern. Two thousand years ago it was an armoured force of footmen that bestrode the world; a thousand years ago it was an armoured force of horsemen; today the world reels before an armoured force of vehicles. There are other connections between protection and mobility; but this is the first and simplest. Armour increases until it becomes as heavy as a man can carry in battle, or a horse can carry, or a vehicle. Then it becomes over-heavy....

The connection between armour-protection and hitting-power is a little more complicated; before we deal with it, let us attempt a division of the history of war into the periods that are indicated by these armoured wave-crests.

First there is a time when men knew little of metal-working, and few men had armour. This pre-history of war need not come into our picture to any great extent, because we know little about it and few clear tendencies are to be seen in that 'dark backward and abyss of time'. This period ends with the battle of Plataea, when the Greek armies cleared the Persians out of Europe. So we get our first period:

I. *First unarmoured period – pre-history to 479 B.C.*

From Plataea onwards the armoured foot-soldier mattered most in warfare, until the Roman legion was destroyed at the battle of Adrianople in A.D. 378. That forms the second period in our history of war:

II. *First armoured period – 479 B.C. to A.D. 378.*

Cavalry then became the main arm that won battles; it was usually a fairly light cavalry, not fully armoured. It fought by missiles more than by shock at close quarters. But armour was coming back again and we can roughly date the return of armour to predominance with Charlemagne's victory at Pavia in 774:

III. *Second unarmoured period – 378 to 774.*

From this time onwards the heavy armoured knight dominated the battlefield. But in the slow-changing Dark Ages and Middle Ages the value of the archer as an auxiliary arm gradually increased, until the Plantagenets found in the Welsh longbow an auxiliary to armour that could master armour. We date the end of this armoured period with the Battle of Crécy in 1346:

IV. *Second armoured period – 774 to 1346.*

Our next period has many tendencies within it; it is the whole period of modern warfare up to the point when the tank becomes important. The development of modern industry produces an immense increase in fire-power, and war becomes long-range work in which you scarcely see your enemy. But with the tank, and with the battle of Cambrai in which the tank was first given its chance, armoured protection comes back into the picture. We therefore date this period:

v. *Third unarmoured period – 1346 to 1917.*

And finally we have the present period, in which the armoured vehicle is the dominant arm in land fighting. We date it:

vi. *Third armoured period – 1917–?*

If we leave out the fringes of the pattern, leave out the pre-historic and the present day, here are solid slices of the past averaging over 500 years each. Can we find definite tendencies in the development of warfare in each of these periods? Can we find definite tendencies, within each of them, that explain their change into the next period?

Yes. Most of the rest of this book consists of studies of the tendencies within each of these periods. But first it is necessary to define more clearly what I am writing about when I use the words 'tendencies in war'. I am not writing about 'principles of war'. Many attempts have been made to draw up a list of constant unchanging principles of war. They have all become out-of-date.

Today, if anyone was rash enough to list a body of 'principles', of unchanging and essential rules of the art of war, his list would probably last a few years, if that. Modern war gathers up into it so many other techniques, depends on and changes so with the changes in those techniques, that it must be thought of as a process in continual development and at the same time almost continually being destroyed.

In September 1939 an important document in the English language listed nine 'principles of war', of which mobility, the power to move, was put ninth. Next year the Nazi blitzkrieg in the Low Countries and northern France proved that the place of mobility in the list was nearer the top than the bottom. (I think in the next similar list it came second or third!)

So I am not trying to establish 'principles of war'. I know the nine listed: I know other lists of the same sort. They have never helped me much. They may be of more use to others; I doubt it. I am quite certain that some 'principles of war', in the forms in which they are usually taught, are dangerous to the modern soldier. Take for example one of the 'principles' laid down in a recent popular book on modern war.

The author of this book, who was at one time one of the editors of the U.S. *Army and Navy Journal*, believes that 'certain scientific principles of war craft ... are always applicable. ... They are as unyielding as the laws of mathematics. Just as a college professor and a kindergarten child both will get the wrong answer if they fail to subtract properly, so an army commander and a squad corporal both will court defeat if they fail to protect their flanks.'*

This is the 'principle' – always applicable and unyielding – given on the first page of the first chapter of this modern work. A few months after this book was published the German army struck in France. Did it 'protect its flanks'?

Since March 1918, when infiltration was first practised as the basic tactics of an army, the idea of security by protection of the flanks has been dropped or altered. It no longer governs tactics in the attack. The idea has a limited value, tactically, in defence – a modern defensive position is organized for all-round defence, and has no 'flanks' in the old sense; to talk of 'flanks' in connection with such a position is misleading, because it presupposes the idea of a linear front, and a known direction from which enemy forces approach.

Strategically, the question of 'protecting your flanks' is also very different from what it was in 1917. Troops manoeuvring by vehicle (tank, lorry, etc.) do not need to protect their flanks against troops shackled to the pace of men marching. Fast-moving troops can carry out their attacks without fear for their flanks, because they go too fast for slow-moving troops to get at those flanks. Later, after the fast-moving troops have broken through, the edges or flanks of the break-through may need to be 'puttied up' with infantry and guns.

Therefore the words 'protect your flanks' do not now apply

*Lowell M. Limpus, *Twentieth Century Warfare*, New York, 1940.

to many decisive operations of modern war; when they do to
some extent apply to such operations, it is in a new sense, a
sense different from that in which the words were used in
1917.

I use this example to show why I reject the idea of unchanging 'principles of war'. But, on the other hand, I repeat
that war has tendencies; that these tendencies can be 'isolated
out' and understood; that the armed forces understanding these
tendencies, and getting ahead in the application of them, will
beat the armed forces that ignore or go against these tendencies.

What are these tendencies in briefest outline? They include
the pendulum swing between armour and projectile that I have
already stated, when dividing the history of war into my six
periods. They include the general increase in mobility that I
have already stated for the armoured periods; foot, horse,
motor-driven vehicle. And closely connected with these primary tendencies is the swing of the pendulum between the development of 'shock weapons', for close-quarter fighting, and
of projectile weapons for long-range fighting. When armour
matters a great deal in warfare, shock weapons are normally
more important than missile or projectile weapons. Armoured
men can get to close grips with their opponents. And the fact
that most fighting is done today with projectile weapons does
not obliterate this tendency; now that we are back in an
armoured period, we still get a change towards close-quarter
fighting, and a change towards weapons such as the tommy-
gun and hand-grenade that are suitable for this form of fighting. Armour and in-fighting are close companions, throughout
warfare.

Though there is a swing of the pendulum between shock
weapons and the projectile weapons, it is usual for one of these
forms of weapons to supplement the other. Sometimes one
type of weapon is used for one part of fighting and another
type of weapon is used for another. You can see this almost
fundamental pattern of warfare in any group of schoolboys
fighting. They begin with stones or shot from catapults, or with

snowballs or with anything they can throw. Then if the sides are fairly evenly matched and it is a real fight, some of them leave off throwing things and come to close quarters with sticks and fists, perhaps with feet. When one side is beaten and scattered, throwing begins again; 'mopping up and pursuit' may continue by 'shock action' but 'rearguard actions' are fought mainly with projectiles. In these small schoolboy battles fighting is still the same in essence as it was at various times before the last century. Throwing things, the use of projectiles, is in this form of warfare a useful accessory but not usually a decisive one. Hitting, at close quarters, is among schoolboys, as it used to be very often among soldiers, the decisive method of fighting.

Later in this history of weapons we shall see how a great change has happened in the past few hundred years, which has led to the use of projectiles becoming the main way of fighting, and has made out-of-date and useless practically all the weapons by which men used to hack or slice or disembowel each other from a distance of a few feet. These weapons have become relics of the past. The sword, still worn by officers in peace-time and by cavalry, is a decoration, and of rather less value on the modern battlefield than the bow and arrow. The spear, that in the ancient world broke empires, and in the Middle Ages was the main weapon of the armoured knight, became in Queen Victoria's time the lance, carried by some cavalry regiments more for the sake of the pretty pennon that fluttered from it than for its use in action. The last important charge by lancers was at Omdurman in 1898. The lance was finally abolished as a weapon of the British Army, by War Office order, in 1928.

The last of these shock weapons, the bayonet, we shall deal with when we come to it. Here we are concerned with the theory of the thing, and this is the important theoretical point already touched on. Even when men fight, as they do today, mainly with projectile weapons, there is still a 'swing of the pendulum' between long-range fighting (such as the artillery battles of 1916–18) and short-range fighting (for example, infantry using hand-grenades against tanks). In 1940 and 1941 shock action was not carried out mainly by shock weapons. Projectile weapons were used – but used at short ranges. Now perhaps the pendulum is swinging back, the range lengthening. This is the sort of thing that study of the past can make comprehensible to us.

The main process of development, during a period when armour and shock weapons matter most, is at first towards the production of an army that can hit like a single heavy hammer. But after a time it is found that such an army is either overloaded with armour and weapons, or is too clumsy to be able to manoeuvre successfully against a light and agile foe. So there comes in a second development, by which the army is split up into a number of small units, each under its own separate commander, which are capable of combined manoeuvring. Thus the Roman legion was a much more subdivided organization than the Greek phalanx which preceded it. And with this tendency towards sub-division, there is an accompanying or subsequent increase in the importance of auxiliary arms or units.

During an armoured period, then, we first get simple integration, the making of a solid; then we get complexity coordinated.

During periods of unarmoured warfare, when missiles and projectiles are relatively more important than the weapons used in hand-to-hand fighting, armies tend to depend more and more upon light and mobile troops. There is a tendency away from close order and towards open order. There is a continual tendency towards the production of weapons that can hit your enemy a long way off, before the enemy's own weapons can do you any damage. (This General Fuller has called 'the constant tactical factor'.) There is another tendency towards the achievement of rapid fire; if the longbow will fire three times as fast as the crossbow, at the same range, it can be more than three times as useful a weapon. These two tendencies have gone on until modern science and industry have made it possible for each man to carry a weapon with which he has a chance to kill any enemy he can see; modern weapons, in the shape of long-range artillery, will also kill at a greater distance than it is possible for the men handling them to see; and some modern weapons can be fired ten times a second.

These fairly simple patterns of the development of warfare become complicated by others. There are the questions of fortifications and of transport. When great cities grew up men learned a lot about building. And this made possible the development of fortifications; there are periods in warfare when the arts of fortification developed so far that they influenced tactics much more than the arts of metal-working and weapon-making. And people moving to and from the cities, and goods being brought to them, made roads necessary, and wheeled transport. The development of roads, and the occurrence of periods of history in which good roads were allowed to fall into decay, has always affected tactics to a very great extent. Wheeled transport has often dominated supply, and therefore the size and mobility of armies. Today it does more than that.

The size of an army never depended on a simple decision by a king or emperor: 'We will have another hundred thousand soldiers.' The size of an army depends first on the agriculture and the general level of production in the country concerned. How big an army can be brought to battle is often limited by the food and other supplies it needs, and by the stores and supply (transport) services available.

The 'political and economic' factors or 'patterns in war' I put down here, for the purposes of my argument, as 'complicating factors'. Some, like those I have mentioned, arise from the obvious connection between the technique of a society, its civilian skills, wealth, knowledge, trade, industry and agriculture, and the fighting forces that are in part developed or conditioned by all these economic factors. Others arise from the moral or political growth of societies.

War is sometimes spoken of or written of as mainly a matter of heroism and ideals, of courage and service. These certainly occur in warfare, and often become more noticeable during war than they are in peaceful civilian life – though it is worth remembering that courage is usually useless against much better weapons. Yet even courage has often a 'sensible' reason for it. The history of a country and its geographical position sometimes make some of the war aims of that country essential to its life and growth. If these war aims fully penetrate the people of that country, and are deeply felt by them, the soldiers who represent that people in battle will be different from, and better than, soldiers who know nothing about what they are fighting for, and have no national interest in the outcome of the struggle. In the same way, men who feel themselves free, and able in some way to take part in the decision whether there shall be war or shall be peace, put more of themselves into any fighting that comes their way, because they feel that it is something they have themselves helped to decide, a responsibility they have chosen. In this and other ways the politics or class structure of a society affect the morale of its soldiers, and therefore their tactics.

Another similar factor has a direct influence on tactics. As men develop new tools and new weapons, new ways of living and of fighting, each war and even each battle can be of a slightly different shape from those preceding it. There are some forms of society that change very little. Asiatic monarchies, ruled by a king or emperor who is supposed to be descended from some god or other, are usually organized on the lines that nothing ought to change; all old customs are good; everything ought to be done just as it always has been done in the past. And more progressive nations sometimes take a breathing spell, when customs and past ways of doing things are preserved by conservatism. When societies of this sort go to war, their generals and other soldiers have an out-of-date idea of what war is like. They do not alter their tactics to make full use of the new weapons that science and industry have made available. They have a reactionary or conservative theory of warfare, or they have no theory of warfare whatever, and just carry on by habit according to regulations laid down long before. Such societies produce armies that are usually destroyed by the armies of nations which are more ready to adopt new methods, more ready to face changes and to learn quickly the use of new things.

Because many people nowadays believe you can measure strength in battle by simply counting the numbers of tanks or divisions available, it is worth emphasizing that questions of technique, of weapons and tactics, have often been much more important in war than the relative size or productive wealth of the countries concerned, or the numbers of the populations involved. A great master of the art of war said that 'God is on the side of the big battalions'. But the remarkable thing about warfare through thirty centuries is how often the big battalions have been defeated, in spite of that advantage, by forces very much smaller. It is only when war falls into a rut, and the armies contending fail to mobilize the forces making for change and progress within their own nations, that battles become a simple matter of counting the heads on each side. Then, in that stagnant and unskilled sort of war, you secure victory because you still have a few men left when mutual slaughter has killed off or exhausted enough of the men opposed to you.

I have chosen to treat many of the ways in which warfare is affected by economics and politics as 'complicating factors' for a simple reason: I want to isolate out from the extremely complex business of battle a small number of tendencies that I believe to be inherent in the nature of war itself. They seem to me tendencies that are part of the very stuff of warfare, and to govern the other 'complicating factors', using or rejecting them. Take for example the tendency that General Fuller calls 'the constant tactical factor'; the tendency already mentioned to produce weapons that kill at longer ranges. It is a tendency that is natural to war, inherent within it. (I agree with General Fuller that it is important, though I am certain there are other tendencies of equal importance that marry with it or conflict with it.) This drive towards the production and use of weapons that will kill your enemy while he is a long way off, 'before he can get at you', governs and employs much of the civilian skills of a nation preparing for war. Today it produces long-range bombers. Two thousand or more years ago it produced longer spears than the world had seen before. Whatever the state of a society, or its level of technique, this is a 'constant'. And yet men in battle today are often at closer grips with their opponents, fight at shorter ranges, than they did twenty-five years ago. This is a paradox that can only be resolved when we see that other 'constant tactical factors' exist, which interact with that mentioned – for example, the tendency already mentioned to increase armour up to the maximum that can be carried effectively, or even beyond that maximum.

The tendencies inherent in warfare usually take soldiers by surprise; as these tendencies develop they take on the horrid appearance of 'dastard tricks' of the enemy, that no one had thought of, tricks or techniques that cannot be resisted. Superior civilians sometimes condemn soldiers as wooden-headed because of this; but foresight is unusually difficult for soldiers because war is a special sort of science in which experiments can seldom be repeated with any exactitude, and in which the discovery of reliable principles for planning and action is unusually difficult. Compare it, for example, with engineering.

The science of engineering changes, because new materials are discovered and new processes. The engineers' products become more powerful or more speedy or more economical, cheaper to make or giving better results; and although there may be pauses in this process there is a fairly continuous line of advance. At the same time many of the fundamental things known about engineering do not change. There are constants. The principle of the lever remains unchanged, although better levers are developed to replace those more primitive. The principle that we were taught at school to call the 'law of conservation of energy' applies to the most modern Diesel engine as it applied to the first steam engine. Each invention or alteration in technique is built up on a pyramid of proved and tested facts discovered in the past. This is a science. War is only in part a science; it contains few constants.

There are thousands of years of warfare which are governed

by simple propositions, such as the proposition that men drilled to co-operate will beat an equal number of men not accustomed to co-operating, or men armoured will beat an equal number of men unarmoured, or men with weapons that can reach a long way will beat an equal number of men with weapons that reach a short way. But always these propositions are complicated, in each separate and 'real' battle, by questions of leadership and morale, by the weariness or the energy of soldiers, the food and rest that these men have had, by accidents of weather and of the countryside. The engineer who has designed and tested an aeroplane motor can watch it wear out in the certainty that he will learn some of its weak points, and then make a better engine; the commander going into an important battle may have no second chance, and can have no effective rehearsal that includes all the facts known and unknown to him. He can move pins or make pencil marks over maps; he can play his war game or train his troops in manoeuvres; but the actual test of battle comes to him each time as an unknown grouping of dangers and opportunities. In his orders there may be clear principles that he believes he is following, but these principles later seem to the impartial observer only very rough approximations to reality. Battle is unrehearsed – even unprincipled. It is, in this sense, an art, and a particularly difficult art.

A group of dancers will practise their ballet in conditions very similar to the conditions under which they will perform; an army practises war under conditions necessarily unreal. The German army at one time is said to have practised at manoeuvres with blank cartridges in which there were a small number of live loaded cartridges, so that real bullets occasionally whistled by, and the army was given a fairly close approach to reality at a small cost in casualties. I have seen the same thing done usefully at a British school of 'battle drill'. The war dances of certain primitive tribes are also on occasion dangerous, and good practice. But most armies practise battle in conditions very different from those in which they must perform; because of this, war is one of the most unstable of the arts. You can rely on dancers or singers to reproduce in their performances, night after night, very nearly the same level of excellence; armies only 'do their stuff' at intervals, always in unrehearsed positions, and usually at unexpected moments.

It is precisely because war is not continuous that war is difficult to understand or to develop logically. (This is not an argument that war should be continuous.) During the intervals between wars there is no effective measuring-stick for old weapons and tactics, or for new. Industry has a measuring-stick, in profit or production; medicine has one, in the death-rate or the number of cases of illness. A soldier in peace has no such guide; and it has seldom been suggested to him that he should take as his aim the full employment in warfare of all the civilian skills and resources and developments of his nation in peace. This brings us to the final point to be made in this chapter.

I have tried to trace some of the tendencies, inherent in war itself, that govern all warfare and divide its history into a number of periods, within which changes different in scale but comparable in type have occurred. Next, I have tried to outline these tendencies within each type of period, armoured or un-armoured. Now I must try to state the causes for the changes from one period to another, the causes for the big breaks or discontinuities or revolutions in the history of war. The first of these causes, inherent in the nature of war itself, is the one we have just been stating: the difficulty of judging during peace the probable shape of the next war.

I have already pointed out that conservative societies base their ideas of how to fight on the past, and ignore the threads of change. Armies have been in the past, in most societies, some of the most conservative elements in those societies – and this is natural for the reasons I have been labouring. Old ways of war – become old. They are successful for a long time; in particular the almost invulnerable armoured man is successful, until it seems he always will be successful. Then the break, the change, comes partly because he has been so successful that he has not felt the need to alter, to keep up with the times, to do new things. That is how old ways of war lose; but how do new ways win? Where are the living roots of that revolutionary sort of change in weapons and tactics, which is different in quality from the gradual developments occurring within each of the periods of war?

I believe that the positive roots of this sort of change lie outside the normal development of warfare. Changes of this sort are not in their essence technical developments inherent within war; they arise more from the 'complicating factors' of economics and politics than from war as a separate science. The change from an armoured to an unarmoured period, or in the reverse direction, occurs normally when the peoples who make up a nation, or several nations, discover for themselves new forms of social organization, or in some other way release new popular energies.

This sort of change, producing a new way of war, often comes because a people express themselves in a democratic or popular or revolutionary way; it always comes because of civilian or non-professional intervention in warfare; it does not come from armies separated by their profession from the rest of the nation. These propositions are challenges. To myself, writing as the Armies of Fascism, which have swallowed up and in a sense become the Fascist States, are still 'on top', these propositions are also hopes.

## 2. Some Talk of Alexander

It is not of great importance for us to study the first un-armoured period in warfare, or to attempt to trace tendencies and developments in that period. In this chapter, therefore, we simply note the principal characteristics of the period, and then go on to the first armoured period, which begins with the defeat of the armies of the Great King of Persia by the fully armoured citizens of Greek city states.

The weapons of the first soldiers were of course the weapons with which men had hunted wild beasts for thousands of years. In the hunting of wild beasts, for food or safety or sport, men learnt to co-operate in groups; and these groups, when men were hunting men, became the first armies or units of armies. Weapons changed slowly; arrows tipped with flakes of flint, such as had been used by cavemen of the farthest past against wolves and deer, were still employed by 'native levies' that formed part of the Persian army invading Greece in 480 B.C. Other troops in this army had wooden javelins with fire-hardened points. Others carried slings with which they threw pebbles of about the weight of a cricket ball. (One army of an earlier date, mentioned in the Bible, had 700 slingers out of 27,000 men; they 'could sling stones at an hairbreadth, and not miss'. For some reason – possibly because they carried swords in their right hands – these Biblical slingers were left-handed.) There were better armed troops in the Persian army of 489 B.C.; the soldiers that we have mentioned were from the more backward parts of the world conquered by the Persians. Their weapons show what all weapons must have been like in the far past before there was any written history.

The armies of this far past must have consisted mainly of

light troops; they were not drilled or ordered; they fought like mobs or herds. And they fought mainly with projectiles. If they came to grips hand-to-hand, the normal weapons would probably be stone axes or wooden clubs, for in those days men did not know how to handle metals, and therefore could not make any effective stabbing instrument. Even when they later had some metals, they could not sharpen them as blades; they used tomahawks, or similar weapons.

Huntsmen and soldiers, in the far past, were always looking for the best sort of rock with which to tip their arrows and throwing-spears. We shall never know who first discovered how to pound up metal-bearing rock and heat it in the fire until the metal melted and could be shaped. But we do know that, quite naturally, the first metals with which men worked and from which they made weapons, tools and ornaments, were the softer metals – gold, silver, copper, etc. Of these, gold and silver are too soft for anything but ornament and coins (though royal troops in some Asiatic monarchies had decorative spears tipped with silver or gold). But from bronze you can make fairly sharp points for arrows and spears, and cutting blades for short swords. The blades have to be rather thick; they blunt easily, and bend fairly easily; it is impossible to make a full-length sword from bronze. About the period when the mists of the past are clearing a little, and we get the first stories and poems that are the beginnings of history, we find that the more progressive peoples are fighting their wars with bronze weapons. And they are beginning to wear bronze armour.

Any sort of metal armour, even if it is made of soft metal, will protect the wearer against light or blunt projectiles that are not travelling very fast. So as far back as the siege of Troy, perhaps a thousand years before Christ, the fully armoured man seems to have been able to advance in spite of most of the projectiles he was likely to meet; he could, if he chose, come to hand-to-hand fighting with his opponent. An army at this time consisted of a relatively small number of great men, who possessed this rare and costly full equipment of armour, and a very much larger number of lightly armed hangers-on whose arrows and throwing spears were still the main weapons; their tactics still dominated the battlefield. Because the fully armed men were so few they often fought in single combat; but at this period they seldom seemed to have moved in groups that had any definite formation to them – though there is a description in Homer's *Iliad* of a line of men standing in order 'like a wall'.

A typical battle of this period, if we are to believe Homer (who although, or because, he was a poet was also a very vivid war correspondent), begins with the armies forming rough lines opposite each other. A Trojan leader 'shakes his two spears in the air' and challenges some Greek to come and fight him; he 'dares the bravest of the Grecian race'.

One of the Greeks takes up this challenge; the Trojan has a fit of nerves and tries to get lost within his own army; but his brother jeers at him, he recovers his nerve, and after an immense amount of palavering it is decided that these two shall settle the whole battle by single combat. They then toss up to decide who shall first throw a spear at the other. After the spears have been thrown and the Trojan champion has been well beaten, another Trojan picks up a bow, made of two goats' horns joined together, aims at the Greek champion, and wings him. Other Trojans, disappointed because their champion has had the worst of it, 'rush tumultuous to war'. So the fighting becomes general; each of the leaders gathers his men; chariots are driven round at speed (one of the Greek leaders tells his charioteers that if one of them gets tipped out of his chariot he should 'mount the next'). But most of the officers mentioned in the subsequent casualty lists are those wounded or killed by spears thrown at them with such force that they penetrate shield or helmet. Arrows account for a few of the killed and wounded, and there are some cases in which men are injured by big rocks thrown at them. The number of those hurt in close hand-to-hand fighting is very small.

The chariots in this case are not often used for shock action; they are simply mobile platforms from which the soldier throws his spear, or from which he dismounts in order to do some real fighting. In other parts of the world, at the period before real history begins, chariots are mentioned as the favourite battle units of the armies of the greater kings (with, usually, cavalry as the other main force in these armies). The Israelites, fleeing out of Egypt, are not afraid of King Pharaoh's archers or spearsmen, but of his chariots and his horsemen. And hundreds of years later when 'the Assyrians came down like a wolf on the fold', it was their chariots that impressed the chronicler. It seems certain that at a fairly early period the chariot began to be used as a shock weapon; it was driven straight at the opposing army and was intended to smash through that army, knocking down or running over any man who was in the way. It was much more useful for this purpose than the cavalry of the period. The horseman, having then only a primitive saddle, and no bit in the horse's mouth, and no stirrups, could not control his horse well enough to make a real charge possible. A horse carrying a rider who has not much control over him will swerve away from the enemy ranks, at the last moment of the charge. But the charioteer, bracing himself against the wall of the chariot, and hauling on the long reins (connected to a bridle, if there is no bit) is able to control his horses by main force and overcome their fear of the obstacle in front of them.

Chariots at all times remained special weapons of opportunity – weapons that can be used rarely, only effective at the right place and moment. Because horses are large targets and can be put out of action by a few arrows, chariots could not be left standing under enemy fire. And because chariots need very smooth ground over which to operate, they have always been weapons more suitable to the desert and the plains than to warfare in broken country. When men had learnt a good deal about the use and tempering of the harder metals it became customary to lash scythe blades to the wheels and axles of the chariots. A Roman poet, Lucretius, describes these arms in a most terrifying way:

They lop off limbs so instantaneously that what has been cut off is seen to quiver on the ground before any pain is felt. One man perceives not that the wheels and devouring scythes have carried off among the horses' feet his left arm, shield and all; another while he presses forward sees not his right arm has dropped from him; a third tries to get up from the ground after he has lost a leg, while the toes of the dying foot quiver on the ground near by.*

But although this poet makes it sound as if the wild chariot was an impossible thing for the ordinary soldier to meet successfully, his countrymen of the legions usually despised them as tricky and unsatisfactory weapons, too dependent on ground to be of much solid value. The Roman legionaries did not often think it worth while to take chariots into action, and usually kept them for triumphal processions.

The main business of battle, many hundreds of years after the siege of Troy, was still in the stage that is dominated by projectiles, and to a lesser extent by cavalry and chariots. As iron and bronze became more common, and men more skilled in working these metals, it became possible for small armies to be fully equipped with armour so effective that few projectiles could penetrate it, and with solid metal-tipped spears that an unarmoured cavalry could not face. It was impossible to equip the enormous armies of the Asiatic despotisms with this complete kit of armour; only the officers could be properly protected. But before 500 B.C. most of the citizens of Greek cities possessed the full equipment of the Hoplites, the Greek heavy infantry. This equipment includes a large helmet covering the back of the neck, and sometimes the cheeks and chin, as well as the top of the head. Most of the Greek helmets were plumed, and the plumes of horse-hair or some other stiff material made the soldiers look tall and frightening – an idea that still persists in the busbies and plumed hats assumed by modern soldiers when forced to wear the fancy dress called full uniform.

The Hoplite wore a heavy breastplate and backplate of metal, joined at the sides and shaped to fit him. These plates were made of one piece and must have been extraordinarily uncomfortable and exhausting in a hot climate. It seems to

*Lucretius, *De Rerum Natura* iii, 650–62.

have been normal for a Greek soldier to be accompanied by a slave who carried his armour and helped him into it when action became necessary. From the waist, or just below it, to the knee was not armoured; from the knee to the ankle the Greeks wore what we might call shin-guards; they called them greaves. And each soldier carried on his left arm a large shield, usually round, with which he tried to catch the point of any arrow or spear that was coming his way.

A soldier so equipped could stand up to most of the projectile weapons that were then known, and could usually keep on advancing until he got within striking distance with his spear. The normal Greek spear was, in 500 B.C., a shaft some eight foot long, made of ash or some other hard wood, and tipped with iron. And the Greek armies, which several times defeated much larger Persian forces about this period, did so mainly by the ability to close with and kill, by spear and short sword, their much more lightly protected enemies.

The Greeks' armour was only one factor in their success. Other factors derive not from the economics of Greek life but from its politics. Two of the things that free men delight in – and the Greeks were free men as compared with their opponents – are dancing and gymnastics. The Greeks developed both of these much further than other peoples. They therefore acquired the habit of controlled movement in close order, which is the essence of drill. One historian describing the decisive Battle of Plataea (479 B.C.) suggests that the Persians must have looked with amazement on the Athenian and Spartan infantry moving down hill to attack them, and must have thought that these strange Greeks were carrying out some new sort of war dance. For each Greek soldier moved in exact line with his neighbour, all their shields on the same level so that they formed an unbroken wall. To do this effectively the Greeks had of course to keep in step; keeping in step was something that the Persian armies, made up of all sorts of subject peoples and only roughly lined up by officers with whips, had never attempted and perhaps had never seen.

The third factor that made the Greeks immensely better

soldiers than most of the Persian army was their morale. There is sometimes held to be an antagonism between the freedom of the individual and efficiency in warfare. Throughout history, governments that were concerned to hide their own inefficiencies or double-dealing have insisted that the citizen ought to sacrifice his freedom to know and to argue, for the sake of unity in war. Actually the Greeks and others of their time proved conclusively that this freedom is part of the moral equipment of a people desiring success in war, so that from their time onwards it began to be recognized that slaves are relatively useless in warfare, and that even tyrannical organizations of society need to foster, for the purposes of warfare, a class of citizens who feel themselves to be free. The maintenance of such a class was part of the policy of the Roman Empire during its years of strength, as it had been of the Roman Republic before it.

The morale of the Greek soldiers, based on and developed by their relative freedom, gave them the necessary courage to advance to hand-to-hand fighting. Any schoolboy who has taken part in some sort of gang battle will realize that it does not require so much courage to throw things, and dodge things that are thrown at you, as it does to dive right into the scrum and start using fists and elbows. In the same way men needed to be specially courageous at this period in warfare for them to fight as the Greeks did.

The Greek way of fighting was new; it was therefore condemned by many advanced professional soldiers of the day as being very barbaric and primitive; a Persian general, used to cavalry and chariots and the swift movement of light troops, is known to history for his words of scorn for the Greeks' artless, slogging massacres – and for the subsequent massacre of his own troops by these primitive methods.

The Greeks had developed their normal shock tactics in the warfare between their own cities. These cities were usually walled, and too strong to take. But a city could be starved out by the army of another city, if the hostile army occupied the fairly small patches of very fertile land that provided the first city with food. Therefore a Greek army could not stay for very long behind fortifications; it had to come out and meet the force that was ravaging its fields.

While there were a large number of sieges in the Greek wars of the fifth century before Christ, these were seldom prolonged for more than one campaigning season. But in 413 B.C. an Athenian army was sent to besiege Syracuse, a city in Sicily that had been originally a Greek colony. In this long siege, of great importance to Athens and to its allies and enemies in Greece, the Athenians were not dealing with a normal small city that would starve if its fields were ravaged. Syracuse was an immense trading centre, with allies and tributaries in many parts of Sicily and trading stations in many ports; and it drew much of its food from across the water. It had large stores and warehouses. The Athenian attempt to ravage the countryside, on which they wasted nearly a year, failed to reduce Syracuse to starvation, because supplies could be brought in from other parts of Sicily and from further abroad. Therefore the Athenians attempted to cut the city off from supplies by building a wall behind it, which would shut it in on the land side, while their navy blockaded it at sea. This was one of the first essays in large-scale position warfare (warfare based on fortifications that are not the walls of a city) of which we know much. The Greeks must, however, have had some experience of this form of warfare; as far back as the Siege of Troy they had built fortifications of a sort to protect their ships; later the Athenians, whose city is separated from the sea by several miles, had built two long walls of fortifications joining the city and its port, so that when Athens was besieged supplies could come in by water.

It was probably this siege of Syracuse, at which the Athenians were defeated just before they finished the wall with which they intended to blockade the city, that led to the first development of artillery – of a sort. Dionysius, the Tyrant of Syracuse, is known to have possessed, a few years after the siege, a considerable number of wooden machines called *balistae* and catapults. The *balistae* threw light stones at a certain distance in a fairly straight line. The catapults lobbed heavier stones up in the air, throwing them a shorter distance. When, very much later, another Dionysius invented a machine that would fire arrows one after another in rapid succession, the ancient world had its equivalents of the field gun, the mortar or howitzer, and the machine-gun.

These machines, made of wood, will be described later. They soon became the dominant weapons in position warfare, in the sieges of cities and the defence of walls. But their influence on mobile warfare was slight. Like the chariots, they were weapons of opportunity that needed smooth ground. They could not travel as fast as cavalry or chariots, or even as fast as a well-trained infantry. They could seldom be integrated into ordinary war.

Such long-range weapons were introduced from Syracuse into Greece about the middle of the fourth century before Christ, and were greeted with the usual lamentations that are caused by new weapons. The Spartan leader Archidemus groaned, when he first saw a catapult drop a lump of rock on a man standing some way off, 'O Hercules, the valour of man is at an end!' But in fact warfare in the open field went on very much as it had always done, and valour was still a necessary (though sometimes an over-rated) military quality.

The most usual formation of a Greek army in the fourth century before Christ consisted of a simple line of spearsmen massed about eight deep. The Greek shield was often between three and four feet in diameter, and it seems probable that this line of spearsmen would march in very close order, so that one man's shield tended to protect to some extent the man to the left of him. There was a tendency of Greek armies to edge towards the right as they moved down to battle, because the shield was always carried on the left arm and it was natural to advance this arm and shoulder. If you walk with your left arm and shoulder forward, you are apt to swing to the right a bit. There was also a natural tendency to dislike being outflanked on the right, because that was your relatively unprotected side, the side you could not cover easily with your shield. But in the early Greek armies there was not much sign of the ability to manoeuvre. Tactics scarcely existed; you lined up, charged at the run, fought hand-to-hand until you had made your enemy run, and then looked round to see if there was any more fighting to be done somewhere else. The only sub-division of the average Greek army was that each city contained a number of tribes; these usually fought together under their own leadership.

The Spartans, the best soldiers in Greece, had a more logical and useful sub-division of their armies into units like our battalions and companies, so many men to each. As the Greeks learned more of warfare, they evolved one fairly simple (but very useful) tactical idea: to hold or 'fix' your enemy with one portion of your force, and then attack him from an unexpected angle with another. But they did not specialize their troops or weapons for this fixing and hitting.

Here, then, is our 'first armoured period' fully formed. What of the tendencies within it?

The first tendency was towards concentration of force to acquire a temporary local superiority at some part of the battle. Such a concentration had occurred before by accident, or by a general's choice of his line of attack. Now for the first time it was planned, organized, before the battle started. Epaminondas, the first general to see that an army must as a rule – not as an exception or by accident – be stronger in one part than another, invented the phalanx.

The phalanx was, at first, a wing or division of the army that was ranged sixteen deep instead of eight deep. Because it was so much heavier than the eight-deep line opposed to it, the phalanx could usually penetrate the enemy lines and therefore throw it into disorder. We shall see the idea of the phalanx coming into warfare at intervals from the days of the armoured foot-soldier to those of the armoured vehicle.

Epaminondas also exploited another tactical device. His main striking force, the phalanx, was usually placed on one wing of his army; the other wing was 'refused'. It was held back from the first stages of action, either by being aligned at an angle away from the enemy or by moving into action more slowly than the phalanx. This weaker wing of the army did not meet the enemy forces opposed to it until these forces had been disarrayed by the impact of the phalanx further along the line. This manoeuvre, planned before action, is very similar to the 'oblique order' with which Frederick the Great won many of his battles. And it has its parallel today in the blitz offensive on a narrow front, the attack on a wider front not occurring until the enemy's positions have been pierced.

In the North of Greece there came to the throne of Macedon a remarkable King, Philip, who carried the shock tactics of the Greeks to their logical conclusion. He formed all his heavy troops sixteen deep, made all of them a phalanx, so that if they met an ordinary Greek army the weight of their mass could smash through the Greek line. And he equipped these men with immense pikes, something altogether solider as well as longer than the spears of the Greeks. The *Sarissa*, as this pike was called was not a throwing spear converted into a shock weapon, as the Greek spear was. The *Sarissa* was twenty-four feet long, and was so carried that eighteen feet of its length stuck out in front of the soldier. The heavy butt behind him helped to balance it, but must have made it very difficult for him to turn or manoeuvre.

King Philip of Macedon trained his soldiers to move in very close order, each man only two feet from the man in front of him. Five of the pikes of the men behind the first rank appeared in front of the first rank man. The other ten spearsmen, massed behind, could not at first use their spears, which they carried slanting upwards. But as soon as a gap appeared, through casualties or the sway of battle, in the front ranks of the Macedonian army, there were plenty more spears to fill this gap and maintain an unbroken frontage of six spears to each width of shield. The bristling mass of this Macedonian phalanx could bear down any opposition that it met; it soon made the Macedonian kingdom the dominant power in Greece.

Here is the first tendency in the 'armoured period': the army like a heavy hammer – simple integration. The next tendency, towards complexity co-ordinated, treads on its heels.

Philip saw that this heavy hedgehog of an army he had invented was a bit clumsy, even for the steady foot-to-foot slamming of a normal Greek battle. He therefore split up his army, making each large unit a separate phalanx and leaving gaps between these units. In order that the flanks of the units should not be too vulnerable, they were sloped back so that spears and shields were at a slanting angle rather than in a straight line across the front. Sometimes each phalanx was wedge-shaped.

Philip's brilliant son, young Alexander, divided this heavy infantry into brigades, and sub-divided these into regiments and companies. He trained his men so that each company, eight deep, could wheel to a flank or face about, which is no easy job when men carrying twenty-four foot pikes are massed together, or two or three companies could be massed behind each other. Normally each regiment marched into action separately, but the gaps between the regiments were filled by lighter troops whose business it was to protect their flanks from cavalry or chariots. These lighter troops could keep position when crossing rough patches of country, which the heavy armed men of the phalanx usually had to march round. When it became possible to close up the phalanx or its brigades for a charge, or when there was a heavy attack by cavalry, these lighter troops would usually be withdrawn to the rear or given space within the protecting walls of the heavy companies.

Alexander added to his heavy troops archers, slingers and javelin men, and a certain amount of cavalry. His heavy cavalry were armoured, and their horses were also protected; this cavalry carried long lances and heavy swords. His lighter cavalry were mainly bowmen and were used as skirmishers.

Alexander also carried with him to battle a certain number of light catapults, and used them as a sort of field artillery. Unusual in his willingness to accept new ideas, he made of the army a testing ground for every sort of weapon that was known to his age, and combined all of them with the heavy troops which were the main force on which he relied. There are few soldiers in the ages following Alexander's of whom as much could be written.

Here then are the auxiliary arms appearing fully, to support the armoured phalanx. It was with an army of 47,000 men, made up of the elements described, that Alexander met the Persian king Darius at Arbela in 331 B.C. Although accounts of the size of Asiatic armies have often to be taken with a grain of salt, it seems clear that the army of Darius was not far from being a million strong. If the size of the Persian army is taken as that of this page, the size of Alexander's army would be equivalent to less than two lines of print.

The Persian forces were not only much stronger numerically, they also possessed two special arms. One was the chariot with scythes on its wheels; the other was a formation of fifteen elephants. Elephants had not previously been used in warfare outside India, and it is probable that very few of the Greek army had ever seen such animals. They must have seemed as terrifying to Alexander's men as the first British tanks did to the German soldiers of 1916. But they proved of very little use in battle. Elephants are wise animals that like to live several hundreds of years. They therefore avoid unpleasant and noisy situations, and are likely to stampede during a battle. Why should they face a line of soldiers glittering with spears and armour, and yelling loudly? It is not clear what happened to the elephants at Arbela, except that they were badly scared. (On another occasion, hundreds of years later, when elephants were used during Hannibal's last great battle, many of the beasts stampeded when they heard the Roman trumpets. They destroyed the order of Hannibal's own African cavalry, making it possible for the weaker Roman cavalry to chase it off the field.)

At Arbela the Persian king had drawn up his immense army at one side of a wide area that had been carefully cleared of rocks and other obstacles. The Persian idea was that they could use this levelled space for the chariots. After a day spent on reconnaissance Alexander moved his army in oblique order diagonally towards the Persian left flank. This movement was so directed that the bulk of his forces were likely to get beyond the levelled space, on which the chariots could operate, before they made contact with the Persian flank. The Persian king was therefore forced to send forward first his cavalry and then his chariots into action considerably ahead of his main line, where they could not be supported by the rest of his troops.

The chariots might have done a great deal of damage to the heavy phalanx if they had been able to attack the Macedonian spearsmen without meeting any other opposition. But as soon as they came rattling over the levelled plain, Alexander's lieutenants sent forward lightly armed troops, at the run, to oppose them. They poured their arrows on the horses and charioteers, dodged the chariots and rang alongside them, slashing at traces and even seizing reins. Few of the chariots got through this mosquito cloud of light troops, and fewer still were under much control. Then as the chariots came to the phalanx, each brigade packed a little more solidly together and opened, between itself and the next brigade, lanes down which the terrified horses could bolt. Some of the chariots went right through the Macedonian infantry without damaging it, and were captured by Alexander's cavalry far in the rear.

Such a battle must have been, for the most part, a mixed and whirling affair of cavalry charges, the disordered open fighting of the light troops, the uncontrollable dash of the chariots, the panic of the elephants. There was scarcely more tactics in all this than in the scrambles round Troy. But there was one element of order and precision that was the basic thing in Alexander's strength: the massed power of the phalanx. It might not have succeeded alone; and it is possible that the majority of the Persian casualties were caused by the spears of the Greek cavalry, not by the spears of the phalanx. But it was the irresistible march of five brigades of the phalanx, turning inwards towards the level ground and the Persian centre as soon as the chariots had spent their force, that broke the Persian line and also broke the nerve of the Persian commander.

For many generations after Arbela the phalanx was unbeatable. But there are few Alexanders in the world and many men who think that drill and discipline are the sole foundation of soldiering. In their hands the phalanx became once again a solid body, homogeneous, the parade-ground general's ideal, but not easy to manoeuvre, not sub-divided into separate units, and not combined with other troops in Alexander's way.

Alexander's successors, who split up his empire between them, tried to make soldiers of inferior quality, even slaves, carry out the tactics of the phalanx. But these soldiers had neither the tradition nor the morale of the Macedonians, and they could only act as clumsy 'hedgehogs' of spearsmen, without mutual support and without the power to combine with other arms. The combination of separate and different arms was becoming, even at this early period, an essential thing in tactics. When it was a simple (or stupid) combination, the units merely added to each other; when it was Alexander's sort of combination, the power of each arm multiplied the power of the others.

Occasionally during this period a phalanx of spearsmen met another phalanx in battle. In this case victory usually went to the luckier side, or to the side that had the more solid soldiers in the more rigid formation. War swung back temporarily from the flexible combinations of Alexander towards the formation of armies that were heavy and clumsy hammers.

But in a campaign carried out by the Macedonians in Italy the phalanx met its master. The tough farmers of central Italy, who had gained experience during a series of wars against their neighbours and against invading Gauls, had produced a rather different method of fighting. They used almost the same weight of armour as the Macedonian heavy troops, and a more effective shield, but they trained their men to manoeuvre rapidly in small units, in such a way that they could come in on the flank of any solid body of spearsmen that opposed them, or split up any charge of horsemen. They relied more upon short swords than upon long spears, because long spears cannot be manoeuvred quickly in a new direction. Their first ranks also used heavy throwing spears. The Roman legions were forming.

At first the Macedonian phalanx, if it held together, could usually drive these Romans off the smoother portions of a battlefield; the lighter Roman formations and shorter spears could not oppose the long *Sarissa* handled by a compact mass of armoured men. But the phalanx could not pursue the Romans or split up their flexible small units, or even drive them out of rougher country, woods or hillsides. And after the phalanx had been in action for a time, some of those forming it were wounded by the throwing spears of the Romans, or wearied by the weight of their own great pikes, and they lost cohesion a little. Then the Romans jumped in with their swords, much handier weapons than the pikes for a 'mix-up'. So King Pyrrhus of Macedonia won a number of victories over the Romans, but each victory cost him more than he gained by it. Since that time any costly and futile victory in war has been known as 'a Pyrrhic victory'.

The Roman legions, when fully formed, were 'heavy troops'; they fought in close order; their shields could be locked together to cover them against projectiles. They relied increasingly on throwing-spears. This combination of shock tactics with short-range projectile weapons gave the Romans a great advantage. They combined the order and discipline of the Greek phalanx with a superior 'fire-power'. The spear (*pilum*) could not be thrown far; but it could wound an enemy spearsman before he could get close enough to use even the longest of pikes.

The Roman spear was heavy because more than half its length consisted of iron. The iron point was integral with an iron shaft over four feet long. Inside this shaft, driven well into it, was a four-foot handle of wood. The whole length was six to seven feet.

Against cavalry and light troops this weapon could be used as an ordinary spear. Against a solid infantry it was thrown. Gradually, as the legions found it necessary 'to hit before the other fellow could', and therefore needed a longer range for their 'projectile', they lightened and shortened the *pilum*; the iron shaft became three foot long, or less, and the wooden shaft was also shortened.

The Roman infantry normally fought in three lines, one behind the other, each line eight deep. At first only the front line had throwing-spears and the second and third lines a longer thrusting spear – a remnant of the old Greek phalanx. Later the first two lines had throwing-spears; then all three lines. The lines were not often continuous; the usual formation was like three lines of black squares on a chessboard, with white squares empty between them; and the units could be even more widely spaced.

A legion of 6,000 men was divided into ten cohorts, and each cohort had three 'maniples', one in the front line, one in the second, and one in the third. It was therefore 'extended in depth'; as reinforcements came from the rear lines, they were men of the same unit as those they were supporting, accustomed to the same officers. This, and the open nature of the chessboard pattern, which allowed movement of the reserve units to a flank or allowed them to fill the gaps in the front line, gave the Roman commander full control over his men. He could spread them to outflank, or concentrate them to break through, the enemy's line. Even Alexander's brigades, regiments and companies were clumsy compared with the Roman legions and their cohorts and maniples.

When a Roman army was robbed of this flexibility, given it by its open formation, it lost much of its fighting strength. Its leaders, therefore, had to be careful of their flanks, because flank attacks on a Roman army could squeeze it too tightly together, and the cohorts would then lose their power to manoeuvre. The usual battle order of a Roman army included a force of cavalry on either flank, whose duty it was to hold off enemy flank attacks at least until the legions had got to close grips with the main body of their opponents.

At Cannae the Romans were defeated by Hannibal in a battle that ever since has been the budding general's dream. Hannibal's army was scarcely four-fifths the strength of the

Romans'; his special weapons – elephants brought with infinite difficulty through Spain, southern France, and across the Alps – were not particularly effective. But his cavalry was powerful. In his order of battle his centre (infantry) was pushed forward. As the Roman legions advanced to the attack, this centre gave ground slowly. The middle of the line gave more than the wings. This refusal to meet in a decisive clash gave the Carthaginian cavalry on the flanks time to drive the weaker Roman cavalry units off the field. Then Hannibal's lighter troops, that had mingled with or followed the Carthaginian cavalry, turned inwards and caught the Roman infantry from each flank. The Roman legions, that had been moving forward into a narrower and narrower space as the enemy's centre retired, were pressed closely together; they could no longer manoeuvre; the men who had used up their throwing-spears could not get new ones or be replaced by others with new ones. The front ranks began to be filled with wounded and weary men who could not get to the rear because of the unbroken, close-packed lines behind them. Then Hannibal's victorious cavalry took them from the rear. The net closed; in a swaying mass the Romans were butchered, until nothing that could be called an army remained.

The Romans learned from that battle the tactic of 'envelopment'. From that time onward it was the normal desire of every good Roman general to carry out a *Cannae* – a double envelopment of the enemy's flanks (as it is today of every German general – how many times the German communiqués in this war have contained the triumphant word *zusammenpressen*, squeezing, to describe successful envelopment!). And the Romans learned many other war-tricks from the Carthaginian people whose wits had been sharpened by trading and the sea.

The strength of the Roman legion was greatly increased during the 'world war' of Rome against Carthage, when the Romans conquered Spain and there developed a new weapon. Spanish iron ore provided at that time, and from that time until modern days, the best metal in the world for the making of weapons. The quality of the metal and the skill of the men who tempered it made it possible for the Spanish sword to be sharper, more pointed, and yet lighter than the weapons previously carried by the Romans and by other soldiers of the period. It was not a long rapier such as that used by fencers today; it was shorter than the swords used by many other troops of the period; but its ability to pierce through the joints of armour, and its handiness due to light weight and good balance, made it the dominant weapon in hand-to-hand fighting for a long time. And it did not blunt or bend easily; most swords of the period were only good for half a dozen hard blows; the Spanish sword was good for years of fighting. This Spanish sword was introduced into the Roman army by Scipio Africanus, the general who achieved the defeat of Carthage.

Carthage was a trading empire. Its power was in its wealth, its ships, its possession of protected harbours throughout a large part of the Mediterranean. It had no considerable class of free citizens from which it could form its armies; most of its free citizens manned its navy. It found, as was always found in the ancient world, that slaves made poor soldiers. In Carthage therefore there developed, probably for the first time in the world, a large-scale mercenary army; most of Hannibal's soldiers fought for pay – not as men forced to fight by national or local compulsion, and not as volunteers whose main motive was to defend or extend the power of their own country.

The mercenaries who followed Hannibal were far freer men than those who were tied by law or custom to follow some ruler or form part of some army. This freedom gave them part of their ingenuity, their craft in war. They were dangerous enemies.

In the social organization of the Roman Republic there remained at first much of the primitive patriarchal organization of society. Only the head of the family could own property; all the young men were supposed to be entirely subject to the orders of the oldest man in the family. But faced by the Carthaginian army, and suffering from its ingenuity and power to manoeuvre, the Romans had to depart from this old and hampering social system, which crushed the initiative of its young men and gave them no incentive to act for themselves. The Roman soldier – then always a citizen – began to be permitted, by custom and then by law, to possess a separate form of property, called the *peculium castrense*, the possessions in the camp. This was the first legal form in which the right of individuals (other than fathers of families) to movable property was recognized.

In this way the Roman legion became a body of citizens with special rights of their own which were in advance of the rights of civilians of the same age. The legionary became economically as free an individual as any in the Roman world. With the pride in himself that came from this freedom, and with the morale of the well-fed, patriotic peasant or artisan, he beat the Carthaginians.

He became then or later, as Rome conquered wider fields, a long-term professional soldier. During his years of service he was subjected to the hard and difficult drilling that is necessary for shock tactics; he was trained to march farther and faster than any army had marched before him; and he carried on the march a heavier load than any tackled by soldiers until 1915 or 1916. Sometimes he had baggage waggons, but often he had to move hundreds of miles with this load on his back. The armour and the weapons were heavy enough; each man had to carry in addition a supply of food and two or more large stakes from which the outer wall of a camp could be constructed. Discipline, marching and carrying power depend enormously on morale. The legionaries' morale was civic – 'civis Romanus sum'; I am a citizen of Rome. His fighting strength therefore depended largely on his position as a free citizen of the world's greatest city. The Eagles of the Legion bore the letters S.P.Q.R.: 'the Senate and the people of Rome'. When the legions ceased to consist mainly of such men they were on their way down hill.

In the next chapter we study the end of this armoured period, the breaking of the legion. But here, looking back from today – when the men of the Panzer divisions and the Russian,

American and British tank forces have not only a special power but a special pride as the élite of the armies – we can look back to the Roman legions and understand them and yet marvel at their achievements. The legion was not always undefeated; it sometimes suffered serious defeat when used by inefficient leaders, or without the necessary cavalry to protect its flanks, or against a more mobile enemy or in unsuitable country. The armies of Rome once had the misfortune to be led by a man who had made his way to power in Rome principally by his wealth; and this man, Crassus, took the legions to utter defeat at the hands of Scythian cavalry who used scarcely any weapon but the bow. These bowmen never fought pitched battles, but harried the marching legions for hundreds of miles across the half-desert spaces into which Crassus unwisely led them. There is a limit to which any army can go against the guerrilla; and the Roman legions found that limit in the deep forests of Germany as well as in the bare uplands of Persia. But these were wars at the edges of the known world; throughout the area that was then civilization (or our civilization, for in this book I have to leave out China, separated from our world by deserts and by mountains and by ignorance) the Roman sword and throwing-spear and armour were almost unconquerable. From the century before the beginning of the Christian era right up to the fourth century A.D. the soldiers who imposed the peace of Rome upon most of the known world were the solid, hard-marching, heavily armoured footmen, the legionaries.

## 3. Knights Were Bold

Between the two great wave crests of armoured warfare of the past, between Caesar's legions pressing across the English Channel and the knights of Richard Cœur de Lion turning their pennons towards the Holy Land, lie many centuries of change in war. The first of these changes is the breakdown of the Roman legion, the ending of our first armoured period. Then comes the second unarmoured period, the great days of light cavalry. Later again comes a heavy armoured cavalry to rule the world in the new shape of Feudalism. These three changes are the subject of this chapter.

Behind the breaking of the Roman legion there were of course political changes extraneous to war, but usually expressing themselves through war. Somewhere far off, among the nomads of the steppes in Central Asia and Siberia, changes of climate or growth of population or the merging of tribes into nations produced a raiding pressure on the 'barbarians' who lived along the frontiers of the Roman Empire. These restless barbarian tribes and nations kept up a heavy pressure on those frontiers from about A.D. 235. And at the same time within the Roman Empire the concentration of uncontrolled power, and other causes, led to continual civil war. In the sixty years after A.D. 235, sixteen emperors of Rome and thirty would-be emperors were assassinated or fell in battle. The citizens were no longer numerous enough to form all the armies needed; 'barbarians' had to be taken into the legions. The legions fought among themselves, against their emperors and to change their emperors. A wild tribe or nation, the Goths, broke through the solid armoured crust of the legions and reached the centre of the empire, where there were few reserves; these Goths ravaged whole provinces. Provinces rose against Rome, or the Eastern Empire fought the West. To meet the strain of almost continuous warfare the legions were pulled to pieces and lost their sense of pride and solidarity as units; one cohort would be in Britain, another in Germany, another might have been snapped up for a march on Rome by some general who saw the chance to become Caesar.

And with these changes on the political field went others of a more military sort. The small garrisons along the frontier, thin-spread and without enough reserves behind them, came to rely on auxiliary weapons and arms more than had ever been the case before. For their own defence they relied more on artillery, 'engines' throwing arrows or great stones; and for attack on raiding barbarians they began to rely more and more on cavalry, because of its mobility.

Out of the experience of countless raids and scuffles, and of pitched battles between legions, it became clear that artillery was far more dangerous to an armoured infantry marching slowly in close order than it was to cavalry riding fast in more open formations. The reliance on artillery for defence therefore increased the tendency towards the use of cavalry for attack.

The legion lost in mobility to some extent by taking with it, into the field or to its temporary camps, the artillery from its main garrison positions. It could no longer march swiftly enough to catch raiding parties of Goths or Germans or Vikings; therefore again there was need for more cavalry, which could do so.

There was no longer a clear distinction between the free citizens of Rome and the men of the conquered provinces; increasing numbers of the latter came into the legions and into the auxiliary forces. These auxiliaries grew in number. Units of the imperial guard, most of them light infantry or cavalry, multiplied and were considered more important troops than the legions.

The 'barbarian' volunteers or mercenaries, in the Roman army, whether enrolled in the legions or employed as cavalry and light troops, were no worse soldiers in battle than the Roman legionaries had been. The legend that they were not so courageous or so skilled − is a legend. But they followed their paymaster; or they went with any usurper who could promise them a chance of success, of loot, of power. They had no longer the citizen's sense of a Roman community and a Roman peace that made the Legions of the Republic and the early Empire relatively proof against the temptations of Caesarism. And the legions were no longer opposed mainly, as earlier legions had been, to forces less well organized and armed than their own. They were mainly opposed, in the civil wars, to other legions like their own, armed with weapons exactly like theirs. That is why they gradually gave up the heavy *pilum*, which although thrown was as near a shock weapon as any projectile could be, and adopted a lighter throwing-spear, to get more range to 'hit the other fellow first'. This change weakened their power to deal with heavy attacks, and particularly with a determined cavalry charge − which had usually been met by the legion using the *pilum* as if it was a pike. The lighter spear could range farther, but was no pike; it was useless except for throwing.

Meanwhile the possible enemies of the Romans became stronger. In the early days of Rome, when Tacitus first described the Germans, they were 'without helmet or armour, with weak shields of wicker-work, and armed only with the javelin'. In other words they were still in a primitive projectile stage of warfare, undisciplined and unable to fight with shock weapons. But after three hundred years of contact with the Roman Empire they had learnt much. Thousands of their men had served as Roman mercenaries. They had traded across the frontiers, and in exchange for furs or amber they had received iron and armour. Most of their rank and file, at the time when they began to break down frontiers of the Roman Empire, had heavy shields bound with iron and a long cutting sword that gave them more reach than the Romans had. (Metals had improved and become more plentiful; but the Roman sword seems to have been the same in A.D. 300 as in 200 B.C.) Some of the Germans had developed a deadly weapon that they handled with particular skill, the *francisca*, a heavy battle-axe or metal tomahawk which, whether swung or thrown, would split the Roman shield and go through Roman armour. Dangerous though these Germans were, the superior discipline and control of the legions could keep them in check until the whole Roman way of fighting was destroyed by the Gothic cavalry at Adrianople in A.D. 378.

Here for the first time cavalry became the shock troops that mattered. Battles had been won by cavalry before, used in combination with other arms; battles had seldom been decided by cavalry almost alone. The Goths, big men on big horses, armed with heavy lances, swept the Roman infantry into a huddled mass where men were so crowded that they stifled, and no man could lift his arm to strike a blow. When the emperor with all his chief officers and forty thousand of his men lay dead on this battlefield, the day of the legion had at last ended and the day of the man on horseback had begun. And for nearly a thousand years from that day the man on horseback ruled warfare.

What were the root causes of this great change in war? We saw in the last chapter that the change to the armoured footman came with the Greeks, who all wore armour and fought as solid masses partly because they were all citizens on an equal footing. The idea of equality, under king or law or aristocracy or representative, that governed the politics of the Greek cities, governed also their military technique. They were perhaps the first well-drilled armies because they were also the first armies of fairly free men. That is not so much of a paradox as it may sound today: men do things better and more easily by their own choice than they do under compulsion. From these qualities of the Greeks came the first 'break' in war; did the second, the 'break' away from armour, come from the same sort of cause?

Yes, the roots of this change were of the same sort, though naturally they were not identical. The decay of Rome, the growth of Caesarism, made the Empire brittle. Outside its

strained defences milled the barbarians, ruled by the rough
primitive unorganized semi-democracy of tribe and clan. These
barbarians learned from Rome. But they did not learn to form
legions of their own, to become an armoured, drilled and close-
knit infantry. They learned how to make the qualities of their
own societies tell in war. They learned how to become a raiding and charging cavalry, that although unarmoured or lightly
armoured could beat shock troops by shock tactics. And besides their shock tactics they had the missile weapons we have
already described.

Soon these missile weapons became all-important.

Oman, in his *History of the Art of War*, describes the process: 'The day of the sword and *pilum* had given place to the
day of the lance and bow.' He quotes Vegetius' *De Re Militare*,
in which that admirer of the legion deplores the abandonment
of armour: 'So our soldiery went out with breast and head
uncovered to meet the Goths; and perished beneath their
missiles. . . . For what can the footman armed with the bow,
without helm or breastplate, and even unable to manage shield
and bow at once, expect to do?'[*]

Oman continues:

It is, of course, ludicrous to suppose that, at a time when the
cavalry were clothing themselves in more complete armour, the infantry were discarding it for mere sloth and feebleness. The real fact
was that the ancient army of mailed legionaries had been tried in
the battlefield and found wanting. . . . Roman military men had
turned their attention to the greater use of missile weapons for the
foot soldiery, and to developing the numbers and efficiency of their
own cavalry.[†]

Within this period, the second unarmoured period, there was
a tendency towards the development of whole armies of
cavalry consisting almost entirely of archers. The Huns under
Attila became so feared that their name lingers still in Europe
as a symbol of ruthless war; they were a people of horsemen

[*]Oman, *A History of the Art of War, The Middle Ages*, pp. 3 and 18.
[†]Oman, p. 19.

who could pour out a rain of arrows as they rode. Belisarius, the great general who held together the empire of Byzantium, ascribed his victories over the Goths to the fact that 'our own Regular Roman horse and our Hunnish Foederati (allies) are all capital horse-bowmen, while the enemy had hardly any knowledge whatever of archery. For the Gothic knights use lance and sword alone, while their bowmen on foot are always drawn to the rear under cover of the heavy squadrons. . . .'*

The horse-archer whose missle was the dominant weapon of the period often wore armour of a sort and his opponent, wielding a shock weapon, the lance, also wore some armour. But this protection was usually light and incomplete. In the case of the first Turkish armies to clash with the soldiers of Byzantium, armour was almost entirely lacking: 'The Turkish hordes consisted of innumerable bands of light horsemen who carried javelin and scimitar, but relied not at all on their armour for victory.'†

Mention of the Turks has already taken us up to or beyond the period when war changes again, the second armoured period. And here it is necessary to stress that dates are very arbitrary things; ways of fighting do not change, throughout the whole world, in the same year. When at Plataea the armoured Greeks destroyed the army of the Great King, there was not an immediate end to all the old ways of fighting, or everywhere a beginning of the new. In the same way there was no complete and immediate reversal of tactics and change in weapons in the year of the battle of Adrianople, A.D. 378. The change in this case, in the parts of Europe and Asia that were then the 'civilized world', took about two centuries, from A.D. 250 to A.D. 450. In other parts of the world – perhaps in fact as 'civilized' or more – such as China, the changes described may have happened at other times, or not happened at all.

But the main thing I am trying to trace is the development of weapons and tactics in Europe and Asia that leads in direct succession to modern war. And here we have to note that during the 'second unarmoured period' the tendency around

*Oman, p. 32.                    †Oman, p. 204.

the centre of civilization was largely towards an army made of horse-archers, or towards one in which the horse-archers played a dominant part. This tendency reached its fullest development (by a sort of time-lag, an overlapping) long after a second armoured way of fighting, on horses, had begun to develop fully in Western Europe.

The army made up of horse-archers reached its fullest development with the Mongols, the armies of Genghiz Khan. Modern American commentators have seen in the tactics and strategy of Genghiz Khan a parallel to the blitzkrieg of the Nazis. There is the same use of treachery and fifth columnists, the same surprise and suddenness in attack, the same strategy of penetration and encirclement, the same sort of mobility. But there the resemblance ends. The tactics of the blitzkrieg are largely tactics of heavy armoured troops, fighting at shorter ranges than those normal in an earlier sort of war. The tactics of the Mongols were those of very light cavalry, fighting at the maximum range possible with the missile weapons of their day.

Before the Mongols appeared from Asia, war had swung back to armour again in Western Europe. If armour is strong enough, the horsemen wearing it can charge through the arrows and get to close grips with their enemies. As early as A.D. 451, the army of the Huns under Attila had been checked and turned back at Chalons by the shock tactics of a heavier cavalry. In A.D. 732 the light horsemen of the Moors, who seldom wore armour, invading France from Spain, were defeated at the battle of Tours by an infantry armed with the sword and by a heavy cavalry, fairly well armoured, which fought mainly with the spear. In each case it seems possible that the invading armies had tired their horses before action. These armies rode animals suitable for the plains of southern Russia or the spaces of North Africa but not heavy enough for the soft fields and tracks of France.

This latter victory encouraged the Franks in their development of armour. Soon a great leader arose among them, most of whose first laws dealt urgently with the question of the accumulation of armour and its 'economical' use. Charlemagne, king of the Franks, forbade all merchants to export armour from the territory he ruled. The heavy penalties enforced show the eagerness with which men sought this key to success in war. In A.D. 814 a chronicler described Charlemagne and his army riding into northern Italy:

Then appeared the iron king, crowned with his iron helm, with sleeves of iron mail on his arms, his broad breast protected by an iron byrnie, an iron lance in his left hand, his right hand free to grasp his unconquered sword. His thighs were guarded with iron mail. ... And his legs, like those of all his host, were protected by iron greaves. His shield was plain iron, with device or colour. And round him and before and behind him rode all his men, armed as nearly like him as they could fashion themselves; so iron filled the field and the ways, and the sun's rays were in every quarter reflected from iron. 'Iron, iron everywhere,' cried in dismay the terrified citizens of Pavia.*

Probably this chronicler was exaggerating. With Charlemagne, the second armoured period is fully begun. But not all of Charlemagne's cavalry can have worn a complete kit of armour. It is unlikely that any except the best equipped of his men had a metal covering to the neck; they would have a metal cap and a thick leather covering over ears and neck. This was the first form of the knight's head-dress, the hauberk. But soon (that is to say, by the next century) the material hanging from the metal cap or 'tin hat' was fine chain mail, iron links interwoven, and this was brought forward to cover the chin and the neck. The lower edge was tucked under or tied over the mail shirt or 'byrnie'. This had originally been as short as an ordinary waistcoat today; it later became as long as a pullover. And then these two pieces of armour were made into one garment. The mailed shirt acquired a hood of mail which was pulled up over the 'tin hat', and covered everything except a small space through which the horseman could look out and breathe.

Although Charlemagne's men cannot have been so well

*Oman, *The Art of War in the Middle Ages*, p. 86.

armoured as later knights were, we can take the date of his battle at Pavia (A.D. 814) as marking the beginning of the second armoured period.

In the tenth century the better armed men began to wear a strong iron bar connected to the helmet and sloping down in front of the nose. This was to guard against blows at the face that would turn aside from the helmet; it probably lasted for over two hundred years, until the great days of knighthood when the helmet with visor covered the whole of the head and face. In the visor, which could be lifted from in front of the eyes and mouth when the knight was not fighting, there were small holes through which he could see and breathe. This type of helmet is that best known to most people who have seen medieval armour either in museums or in pictures.

In Charlemagne's day the average armoured soldier seldom had protection for his thighs. But a hundred years later the mailed shirt was usually lengthened till it reached the calves. It was, of course, divided to make riding possible. Breeches of mail, which must have been torture to ride in, were not used until much later.

The round shield had almost disappeared. It was still carried by Danes and by some other foot-soldiers. But for the man on horseback it is inconvenient to have a large shield, since the left hand must be used for the reins when the right hand is using lance or sword. A very small round shield, carried, as a rider must carry it, strapped above the elbow of the left arm, only covers a part of his body that is already protected by his mail shirt; it does not cover the more vulnerable lower parts which on a horseman may be exposed to spear-thrust and sword-slash from the ground. The shield of the early Middle Ages was therefore shaped like the commonest form of kite that children fly; the long lower end could be carried to cover the left knee or set across the saddle.

While armour increased in this way, until its weight came near the limit that a strong horse could carry, weapons altered less. Danes and Franks, as infantry, used a double-handed axe, sometimes also double-headed, to cleave through armour. The sword grew longer; but in the early period of the Middle Ages the horseman's lance was seldom a long and heavy weapon; at Hastings the Norman knights sometimes threw their lances as well as thrusting with them. Men were still learning to make good saddles and stirrups; without these a heavy lance is dangerous – if the knight sticks such a lance well into his opponent, the horse goes on but the victorious knight slides off over his beast's tail.

The armour of the Franks was too stout to be pierced easily by the ordinary arrow. And thanks to this armour Charlemagne achieved something that the Romans could never achieve and no subsequent ruler – save for a short time Napoleon and Hitler – could so much as attempt. He united in one state France and Germany. When that unified state fell to pieces at his death, the wars of modern Europe were beginning to take shape.

At the battle of Hastings, which we can take as a convenient point where the early development of feudalism is shown in its completeness, the English represented the more primitive, 'backward' form of warfare of the period, lacking most of this feudal development. They were foot-soldiers; they were lightly armed, their weapons were swords, javelins, clubs, axes and a few short bows. They fought in a solid mass, six or twelve deep, behind a low ditch and a stockade of brushwood and hurdles, roughly put together and covered by shields. Behind their shield wall the English crouched for protection from the Norman arrows, standing up to throw their axes or other hand projectiles when the Norman cavalry came near.

The Norman army was a very different organization. It did not consist, as the English army did, of men raised by a primitive 'conscription', under which every family owning so many acres of land must send to war one man with infantry equipment, or if they owned less a man with some primitive weapon. Duke William of Normandy brought with him most of the barons from that province, and a large number of feudal adventurers seeking 'some for land and some for pence'. Their equipment was that of the foremost knighthood of their day.

His army, however, did not consist only of heavy cavalry. The first line of each of his divisions consisted of archers with a smaller number of cross-bowmen; the second was his foot soldiers – younger sons, knight's squires, poverty-stricken adventurers – most of whom wore mail shirts, unlike the bulk of the English infantry, who came straight from the plough in the leather jerkins they normally wore.

Behind his secondary troops came William's main force, the armoured knighthood.

William opened the battle with his bowmen, to whom the English could scarcely reply. But when these archers came forward within range of the English hand-thrown axes and javelins, they got too heavy a shower of these and broke back to the protection of the Norman infantry, moving forward with spear and sword. This infantry drove back, to the breast-work, some of the English who had been running a little way down the hill to get close enough to fling their missiles at the archers. But against the main line of the English infantry, the Norman infantry – much outnumbered – could not make an impression. They were probably already retreating a little from the missiles thrown at them, and from the English pikes, when William's horsemen came up the slope as hard as they could ride. At many points these horsemen must have swept down the English hurdles; but they could not push through the stubborn English infantry, whose axes cut through shield and mail, felling horses and lopping off limbs. After a fierce medley William's centre and right wing fell back a little, to regain its order; his left wing fled down the hill.

Whether or not this flight was a deliberate manoeuvre, it led directly to the defeat of the English. They had been tormented by arrows to which they could give little answer; they had met and checked easily the Norman infantry. Now after heavy hand-to-hand fighting the English on the western end of the battlefield saw their opponents riding away down hill in panic retreat; saw these horsemen clumsily break into and sweep away the units of Norman infantry that had rallied behind their advance. It was natural for this primitive English army, with no considerable leader save its king to control it, to dash in pursuit after the beaten enemy.

But William was not beaten. He had rallied and reformed his centre. He saw to his left the English pouring down hill, having lost formation and lost the protection of their stockade and their shield-wall. William turned his horsemen and crashed into them as they moved. His heavy horsemen were no longer riding up hill, but along the slope of the hill parallel to the crest. His opponents were no longer stubborn ranks twelve deep, but a scattered line of men running across his front. When Duke William had finished this charge nearly a third of the English army was cut to pieces.

The remaining two-thirds had not moved. William ordered a second general attack, in which the breastwork was probably destroyed entirely, and the shallow ditch filled up with the bodies of men and horses. This second assault fared better than the first, the two brothers of the English king were killed and the English suffered severe loss. But their long line of shields and axes still held the hill crest.

Now William showed the value of the horsed soldiers' mobility, power to manoeuvre. He ordered a large portion of his command to wheel about and pretend to fly. The English thought again that their enemy was broken, and again a large body swept forward from the English line. Again William charged to the flank, and again his cavalry rode over the scattered English ploughmen.

The centre of King Harold's army, his own personal forces, better armed and more solidly controlled than the levies of the English shires, still held their ground. They continued to fight for some hours, under conditions steadily becoming more desperate. William's cavalry roved round the compact mass into which the English line had settled. William combined a series of charges, which kept the English pressed close together, with flights of arrows that could not miss their target. After some hours of this unrelenting pressure the English ranks were filled with wounded men who could not press to the rear because of the close-packed mass of their comrades behind them. The

English had no more javelins or axes to throw, no weapon with which they could reply to the bitter arrows. 'A strange manner of battle,' says a chronicler, 'where the one side works by constant motion and ceaseless charges, while the other can but endure passively as it stands fixed to the sod. The Norman arrow and sword worked on: in the English ranks the only movement was the dropping of the dead: the living stood motionless.'*

In spite of this merciless harrying the English held till evening. Then as a last resort William ordered his archers to fire their arrows high in the air so that they fell from above on the English, some of whom must have been too weary to lift their shields above their heads, some of whom may have been too closely packed together to do this. One of these arrows wounded King Harold mortally; a last charge of the Normans broke through to the king's standards and cut them down.

The battle was won; but even after the winning the English took some revenge. The slope behind the crest on which they had been standing was very steep; many of William's knights charged down it in pursuit of the scattering English and came to grief at a small brook with sharp banks below. It was nearly dark, and neither horse nor armoured man could keep his feet. Turning back, a remnant of the English caught these knights scrambling with their heavy gear along this brook, and killed them at close quarters. This savage counter-attack, which the Normans thought must have been delivered by fresh English reinforcements, nearly caused a panic among the victors. One of William's chief subordinates urged him to retreat before these new forces of the enemy. But Duke William was a better soldier than that; he rallied his cavalry and moved it forward along safer slopes. The English, almost all their leaders killed, and perhaps two-thirds of their army, slipped away into the woods.

The armoured cavalry that won the day at Hastings, and dominated European warfare for a long period before and after that battle, was seldom used in such a way that its

*William of Poictiers, quoted from Oman's *History of the Art of War*.

*strategic* mobility was of importance. Its job was not to ride round the opposing army and outflank it, nor to move swiftly upon some outlying section of it and deal a decisive blow before this detachment could be reinforced. The art of war knew little, at this time, of 'fixing and hitting', or of envelopment. The main job of the medieval knighthood was to act as shock troops, breaking into and riding down the mass of the opposing army.

At the same time it possessed a capacity for *tactical* manoeuvre which was greater than that of any infantry force. William's charges to the flank, by which he caught and destroyed the sections of the English army that rashly left their stockade, are examples of this mobility.

The heavy cavalry, the dominant arm, was used alone or almost alone by some of the less progressive commanders of the period. The use of archers to disturb and harass the array of an opposing army, and the use of infantry to hold positions, to clear difficult bits of country where the knights could not easily make progress, to prepare the way for knights in assault and to form a solid rallying point for them if their charge failed – the use of these auxiliaries in proper combination with the heavy cavalry became the highest level of feudal generalship. William the First, in his handling of his archers, and particularly in his use of high angle fire against a tight-packed target, showed exceptional realization of the value of combining auxiliary arms with his principal formation. The history of feudal war is full of the failure to achieve such a combination; over and over again, during the Crusades, the knights of Western Europe were defeated because their leaders relied too exclusively on the men on horseback. But there were few cases when the man on foot was more than an auxiliary – useful, perhaps indispensable, but not the main force in battle.

In feudal society, in its classic form, all land except that held by the church – and even some of the property of the church – was supposed to be held from the king. The king, in return for his grant of land, was entitled to ask from the men to whom he granted it various forms of service, including military service. If the estate was a large one the service would not only be that of the man to whom it was granted, or some individual of that family, but would be the service of retainers and tenants who in turn were bound by feudal law and custom to act as soldiers, either for a customary period or when required, under their lord.

Feudal society arose because of the need for some security, authority, protection, during the centuries of raiding and looting that followed the breakdown of the Roman peace. To any community exposed to the raids of Vikings or Vandals there was little protection to be gained from any far-off emperor; the strong walls of a local lord's castle were their best shelter, and his power to raise a local force their only hope of successful action against the raider. It was natural that such a society, based on these needs, should concentrate most of the available weapons and armour in the hands of the lord whose social function it became to give some measure of security to 'his' people.

The first thing that the lord saw to was that he himself and his men-at-arms were fully equipped; this was a necessity not only for the protection but also for the governing of his area. He was a professional, his sport or profession was warfare. Other people scarcely needed arms, except for the few who accompanied him to battle. This order of society was particularly suited to the production of relatively small armies consisting of very heavily armoured cavalry. And this 'chivalry' was for a time almost invulnerable to most of the normal weapons of foot-soldiers.

A lasting memorial to this feudal civilization consists of its castles. But with the development of these, and of the machines used against them, we can deal in a later chapter.

Though we have taken the conquest of Britain by this armoured cavalry as an example convenient for our history, it achieved other and greater conquests. While the Normans were winning for themselves baronies in England, the chivalry of Castille was winning Spain back from the Moors. The Moors and Turks and Saracens remained in the older previous way of war: they were mainly lightly armoured horsemen fighting by missile weapons. That is why the armoured chivalry of Europe was able to clear Spain and to invade Africa and Asia during the Crusades.

But the second armoured period, like the first, had within it the seeds of its own destruction. In our account of Hastings we have mentioned the value of the archer to the shock troops, the cavalry. It was the development of this auxiliary that gave the knight his deathblow, in a change that was also the ending of Feudalism and the beginning of the modern world.

# 4. Yeomen Bowmen

'So the knights in the first French battle fell, slain or sore stricken, almost without seeing the men who slew them.'

In these words a chronicler described the fate of armoured knights at the hands of English and Welsh archers in the battle of Crécy, 1346, that ends our second armoured period. We are entering the time when men are killed by those whom they cannot see, the modern age of long-range fighting, which contains so much of the modern world that it must fill several of our chapters.

The bow and arrow had been known for many thousands of years; they come into the first battles of which we know anything. The change that made the bow, from an auxiliary, the main weapon in warfare during the second unarmoured period has already been described. It was retained as a principal armament by the Asiatic 'horde' armies that persisted in that unarmoured way of war, and never achieved the armoured knighthood of Western Europe. But in Western Europe it disappeared for a time, because it was almost useless against good armour. Instead of it the crossbow was developed.

The Romans of the period of the legions had possessed a weapon very like the crossbow, but had never considered it very important. They seem to have left it in the hands of light troops and auxiliaries. It should perhaps be considered the most mobile form of the 'artillery' with which the Romans equipped themselves towards the end of the period of the legions. Then in the Dark Ages it seems to have been forgotten.

In the second unarmoured period, in Western Europe and in Asia, it was not used; this was a period of the horse-archer, and the crossbow was too heavy and clumsy for a cavalryman to use. Somewhere in the tenth or eleventh century the idea of a small 'hand-machine' that could be used by a single foot-soldier was revived. It could fire a heavier projectile, of some use against armour; that is probably why it reappeared at this time.

The crossbow consisted essentially of two short arms of wood and a stock roughly shaped; the arms were much stiffer and stronger than those of an ordinary bow, and could only be pulled back for a few inches by a man hauling on the cord. To 'load' the crossbow, the end of the stock was placed on the ground, and the archer used all his weight to press the middle of the cord down towards the end of the stock, where it was fixed in a notch. The arrow or bolt fitted in a groove in the stock, and the crossbow was fired by pressing the string up, from the notch in which it had been caught, to propel the missle along this groove.

In a later form of the crossbow, called the arbalest, an even greater tension of the string was secured by using a primitive sort of ratchet and a handle, with which to draw back the string to the furthest possible point. It was still necessary to put the stock on the ground, in order that the crossbowman might bear heavily on the ratchet as he wound up the string. This weapon therefore would not fire as fast as the ordinary bow.

The arbalest was one of the principal weapons of the Crusaders. Richard Cœur de Lion was particularly fond of it, and spent a large part of his income on mercenaries skilled in its use.

The ordinary bow, four or five feet long, became of so little importance that in 1181, when there was in England an 'Assize of Arms' (a mustering of all the feudal levies to see that they possessed the equipment and weapons considered necessary), the bow was not mentioned as a weapon. But the following year, at Abergavenny, Welsh arrows penetrated an oak door four inches thick. And another Welsh arrow went through the skirt of a knight's mailed shirt, through his mailed breeches, his thigh, the wooden board that formed part of his saddle, and came to rest pinning knight and saddle firmly to his horse. These arrows were shot from long bows of elm, almost twice the size of ordinary bows. The English chronicler of this Welsh war described these bows as 'ugly unfinished-looking weapons, but astonishingly stiff, large and strong'.

For another hundred years the conservative English kings believed that the crossbow was a better weapon than the longbow. They employed Welsh archers occasionally, but probably these Welsh archers had no particular reason to wish to fight for English kings, and were not therefore very reliable troops. In 1281, we know from the pay sheets of a castle held by the English in Wales, a crossbowman received four pence per day, a longbowman received two pence.*

During their long scrambling campaigns against fiery Welsh chiefs the English learnt to use the longbow; but it scarcely developed into an important weapon until the first Plantagenet king, Edward I, began the conquest of Scotland. The first invasion of this country was an easy matter; the Scots had too few armoured knights to stand up to the English army, and many of these knights chose to be on the English side.

A little later came a popular rebellion led by Wallace, an outlawed and obscure knight from the moors of Galloway. He won the whole of the Lowlands from the stupid sixty-year-old general whom Edward had left behind to keep control of Scotland. Next year, at Falkirk, Edward destroyed Wallace's army, which was almost entirely infantry, and was deserted by its few noble horsemen, who even in those days did not like popular rebellions. To destroy the massed Scottish pikemen, Edward did not rely on his knights alone; when their first charge was checked he brought forward his bowmen to pour their arrows into the Scottish ranks. The Scots could not reply; they had neither archers nor cavalry with which to hit back; their infantry dared not break ranks to go after the English archers; it was the last stage of Hastings over again. This weakness of

*These were the official rates of pay, allowed to the chamberlain of the castle. But chamberlains did not always, in those days, act with complete altruism. If these soldiers got half their nominal wages, they were lucky.

infantry against the combination of heavy cavalry and archers was noted by a Scottish knight called Robert Bruce, who was present at the battle on the English side.

The Scottish revolt went on. Bruce joined it, became its leader and king of Scotland. He fought a guerrilla war – the resource of national and popular movements in all ages. He advised his men to fight 'by hyll and mosse', in the woods rather than in castles, by night surprises and ambushes; he 'scorched the earth', burning crops to starve out the English armies, and gradually cleared the whole country until only three towns were in English hands.

Edward I was dead, but in 1314 his half-baked son marched north to deal with the Bruce. The armies met at Bannockburn. Bruce had chosen a position guarded by woods and marshes so that only a very narrow front was open to the English attack. He was facing odds of two or three to one, and he had to rely on infantry against the heavy cavalry of the English. But the whole tactical lesson of the Welsh and Scottish wars – the value of the longbow when combined with troops capable of shock tactics – was ignored by the English army. Edward did not put archers in the gaps between formations of knights; the archers were set behind them. And the first line of the English knights, when their charge had failed and they could make no further progress uphill against the Scottish spears, were too stupid and too courageous to withdraw and try another charge. They did not even give room for their own second line of knights; they just piled up against the spears and went on hacking and hewing at them, while the second and third English lines waited on the slopes below without a chance of going into action.

Then the forgotten archers were remembered. But they could not be brought close to the Scots, because of all the un-controllable mass of gentlemen packed into the narrow space in front of them. The archers were told to fire their arrows high in the air, and to try to reach the Scots over the heads of their own knights. The arrows hit far more English in the back than Scots in the breast. Eventually the English king fled, with a large bodyguard; left dead on the field of Bannockburn was a larger proportion of the English baronage than was slain in any battle before or after that time.

It was from this fighting in Scotland, and particularly from the casualties of Bannockburn, that the English army learnt that a judicious use of the longbow was essential in battle, and could make all the difference between victory and defeat. The application of this lesson at Crécy, thirty-two years later, began the destruction of feudalism, of the armoured knight, and of a whole period of war – the 'second armoured period'.

The campaign that led to Crécy was a typical feudal raid, not so much aimed at decisive conquest as at the looting of French cities and at the lucrative business of capturing noble prisoners in battle, prisoners whose ransom built many fortunes in those days. This raid, led by Edward III, almost reached the walls of Paris. Edward then beat a hasty retreat in face of the much larger army that the French king brought against him. The crossing of the river Somme delayed Edward so long that the slow-moving feudal levies of the French were on his heels immediately after he had got across the river. Edward, outnumbered by at least three to one, chose a strong defensive position on slightly rolling country where his archers could use their weapons to the best advantage.

Edward's army was, for a feudal army, very weak in armoured knights and very strong in archers. He probably had less than four thousand armoured men against the French twelve thousand or more; but he had over eleven thousand archers, and five thousand Welsh infantry some of whom must have been archers, as against six thousand crossbowmen in French pay and a scrambling mass of roughly armed French infantry, among whom there can have been only a few bowmen.

The English archers were drawn up in a formation that they had learnt in the Scottish and Welsh wars – a chessboard arrangement, often compared by the chroniclers to the way in which the teeth of a harrow are set. It was a relatively open formation, so that each man had room to see and fire through the gaps separating the men beside him or in front of him.

Many of the archers seem to have been served by an aide, who carried equipment on the march and supplied them with a large reserve of arrows during battle. While waiting for the French, the archers, and their aides, dug in front of them small holes, a foot deep and a foot square, in which it was hoped that the horses of the French knights would stumble when they charged.

These archers were not strung out across the whole front of the position, but were concentrated on each wing, their lines sloping a little forward so that they faced inward; while between the two main 'battles' of men-at-arms that formed the English front line there was also a strong group of archers, behind whom were placed the infantry spearsmen. Further back were the king's mounted reserves.

When the king of France received news that the English army was in sight, his own forces were spread out over ten or twelve miles of road from front to rear. He tried to check the advanced troops, but his feudal lords were quite without any idea of discipline: 'None of them would turn back for each wished to be first in the field. The van halted, but those behind them kept riding forward, saying that they would get as far to the front as their fellows, and that from mere pride and jealousy. When the van saw the others pushing on, they would not be left behind, and without order or arrangement they pressed forward until they came in sight of the English.' That is Froissart's account of the beginning of the battle.

With the French king was the blind king of Bohemia and his son Charles, who called himself king of the Romans; there was the wild king of Majorca and a crowd of great gentlemen who had never taken orders in their lives. The only element of order and discipline in the French army were the six thousand Genoese crossbowmen. These mercenaries moved to the front, dressed their lines and began the action with a volley that fell short; firing uphill with fairly heavy projectiles, they had miscalculated the range.

The English replied with their longbows. Their weapon was more accurate as well as more powerful than the bows of previous centuries, and although it threw a slightly lighter shaft than the crossbow, it could be fired at least three times as fast and hit as hard or harder, having a higher 'muzzle velocity'. It looked as if a snowstorm was beating on the Genoese, as the grey goose feathers caught the evening sun and the arrows pierced the leather surcoats and metal helmets of the crossbowmen.

The little ledge on which the English stood was only about 150 feet above the shallow valley in front of them. But the ground curved in a way that was particularly convenient for projectile weapons. Looked at from the English position, most of the ground slopes at first very slightly; then the slope increases gradually; then it flattens out. A projectile, an arrow, falling more and more steeply as its speed fell off, would follow the upper slope, keeping about the same height above the ground, to the flattening out. The archers were using 'grazing fire'.

From their ledge above the valley the English archers poured an unceasing rain of missiles upon the Genoese mercenaries, who were occupied with the complicated process of winding and ratcheting their crossbows. Disorder broke out among the Genoese, which so infuriated the mounted French knights waiting behind their bowmen's line that they attempted to retrieve this first check by an immediate charge, riding down a fair proportion of their own mercenaries. Some of these armoured horsemen managed to get through to the English right, but the effect of the arrow-fire told so heavily on their numbers that their charge achieved little but the unhorsing of the youthful Prince of Wales. All the front ranks of the French horsemen were broken, the animals fallen or stampeding about, riderless. Many an armoured knight was trampled, helpless, on the ground.

The French Cavalry did not realize that the English archers had won the battle. Again and again they charged as if sheer weight of horses and men could win home, in spite of the confusion and the continuous deadly fire of the longbows. As each charge came it met the missile defence and was broken, adding to the welter of dead and maddened mounts, of fallen men and knights. One charge reached one end of the English line, where it encountered and failed to pass a barricade of wagons. Finally, the king of France himself led a great body to the centre, but neither he nor any of his lieges could force his horse against the unremitting fire of the English infantry. Well into the hours of darkness the archers stood in unbroken line, steadily supplied with ammunition and easily making up their few losses.

The remaining French knights rode away. The English line moved forward to loot and to look over the faces of the dead. On Saturday night, 26 August 1346, the battle of Crécy ended. One thousand five hundred French knights had fallen.

The victory of Crécy, and a later victory at Poictiers, made the fighting reputation of the English on the continent. In previous times continental military leaders had looked down upon the English as poor soldiers. 'In my youth,' wrote the Italian, Petrarch, 'the Britons that they call Angles or English passed for the most timid of barbarian nations; now they are a most warlike nation. She (England) has overturned the ancient military glory of France by victories so numerous and so un-expected that they who lately were inferior to the miserable Scots, besides the lamentable and undeserved catastrophe that I cannot think of without sighing, have so upset things by their mettle and by that of the whole kingdom, that I, who last crossed it on my business, had difficulty in persuading myself that it was the same country that I had formerly seen.'*

It looks to me as if the second armoured period began be-cause the Franks of Charlemagne's day were far from what was then the centre of civilization (Byzantium and the lands round it and the empire of Mahomet's followers). Isolated to some extent in their end of Europe, these Franks became the first modern nation-state, uniting the qualities taught by long Roman occupation to those of the 'barbarians' who had broken Rome. And perhaps in the same way the second armoured period ended because the Welsh and English, even more iso-lated beyond the narrow seas, learned a secret or a sport of

*Quoted by C. Field, *Old Times under Arms* (1939), p. 5.

their own, that made the bow a far better weapon than it had ever been before.

The best-trained horse-archers of Byzantium had learned that the bow cord must be drawn to the ear, not to the breast or shoulder. When the cord is drawn to the ear, the eye can look along the arrow. Saracen archers learned this also; but horse-archers' bows are short, and when armour became strong enough to resist most arrows, archery fell into disesteem, and this way of aiming was forgotten. A horseman could not carry a longbow; nor could he easily use his whole strength, pulling a bow, when mounted.

The Welsh and English rediscovered the secret of accurate aim; and they used the heaviest longbow known, practising with it until they could drive their arrows accurately at the less well-armoured parts of their targets. Their arrows could often pierce normal armour, not only because of the high 'muzzle velocity' of the missiles but because armour had become too 'refined'. It had gradually become more and more complete, covering the whole of a man and much of his horse; but this completeness was gained at the expense of a lightening of the material. A horse can only carry so much weight; if the armour is spread out, to cover all the man and much of the horse, it has to be made thinner. Chain-mail was becoming delicate. In much of the knight's armour it was replaced by metal plates, that might be stronger in themselves than chain-mail, but could have dangerous gaps between them.

The English archers scored particularly heavily against the horses of the opposing forces, despite the fact that horses were now more often protected than they had been. Though we have little good information, it seems very probable that the heavy cavalry these archers defeated were so loaded with armour that they could not go faster than a slow trot. They could no longer charge at the gallop, as heavy cavalry had done at previous times. While this is uncertain, it is quite clear that armour had become so heavy that any knight unhorsed was likely to be useless for further fighting.

A knight far-weltered on his back might be unable to get up, because of the stiffness and weight of his kit. A knight coming a 'purler' would hit the ground so massively that he was 'out' from shock for some time. The pitiful inadequacy of the knight off his horse, very like the inadequacy of some sea-beast stranded, was shown at the next important encounter after Crécy, that of Poictiers (1356).

In this battle almost all the French knights rode to the field of action, but then dismounted and tried to fight on foot. Because they moved so ponderously slowly on foot, each division or 'battle' of their army came into action alone, after the one before it had been beaten. And many of these knights seem to have been physically exhausted by their march before they struck a blow.

At the smaller battle of Auray (1364) both sides dismounted. The side that won had kept a small reserve of two hundred mounted men-at-arms, who seem to have ridden their horses when moving tactically, then to have dismounted to fight. Oman writes of this action that the commander of this reserve force, Calverley, made his men 'strip off their cuissarts (thigh-pieces) to allow them to move about more easily – a proof that the full knightly armour had now grown heavy enough to make all motion difficult when the wearer had been wearied by long fighting. Without this expedient his reserve would not have been movable enough for use at each point of the line, as it was successively in danger of being broken through.*

With these combats the man on horseback no longer ruled warfare. And while he was being defeated by the longbow in France, he was falling to other weapons in Switzerland, where the burghers, in their fight to free themselves from the German feudal lords, had formed their leagues and cantons. During the fourteenth and fifteenth centuries they fought frequent battles against the German Empire and the Burgundians. The Swiss fighters were plain townsmen and cowherds, 'ill-conditioned, rough and bad peasant folk, in whom there is found no virtue, no noble blood, and no moderation', as the Emperor Maximi-

*Oman, p. 636.

lian somewhat petulantly asserted. But despite these social shortcomings they proved themselves to be some of the toughest fighters in Europe.

Finding that their spears were of little use against the heavier lances of the knights, the Swiss contrived a new weapon, the halberd. This was a combination of axe and spear, having a sharp point for thrusting at a knight, a heavy axe for smashing down on his helmeted head, and a curved hook for yanking him off his charger. This halberd was the first special weapon of democracy. It gave the Swiss cantons their independence and the Swiss people their custom – still in force – by which each citizen keeps his own weapon in his own home. The halberd became the weapon carried by royal guards throughout Western Europe, because it was a weapon that helped to destroy the power of the feudal barons and therefore made royal power real.

Though the leaders of the English and Swiss at this period were nobles, the relationship of the men to their leaders was not entirely that of feudal allegiance. More often it was a far more modern relationship: that of cash. More and more, the feudal levies were being superseded by archers and halberdiers who were hired professional soldiers, who fought for a wage and looked drily upon the outdated and high-falutin' notions of glory still exploited by their overlords.

Then came gunpowder to complete the process of remaking the ways of war, and gradually to end the use of armour. The use of powder was discovered, in Europe, in the thirteenth century, some say by Roger Bacon. The first 'musket' or 'handgun' appeared in the fifteenth century. Gunpowder used in larger weapons blasted the foundations of that stronghold of feudalism, the castle. Not only could the knight be put out of battle by a musket or cannon ball, before he could strike a blow, but the feudal lord and his people were no longer secure behind their fortifications. Gunpowder was a leveller. It made the townsman, the merchant, the craftsman or apprentice, and the peasant and professional soldier equal in combat with the seigneur. The days of chivalry were passing fast, and many were those among the knighthood who felt the yearning of Don Quixote:

Blessed be those happy ages that were strangers to the dreadful fury of these devilish instruments of artillery, whose inventor I am satisfied is now in Hell, receiving the reward of his cursed invention, which is the cause that very often a cowardly base hand takes away the life of the bravest gentleman, and that in the midst of that vigour and resolution which animates and inflames the bold, a chance bullet (shot perhaps by one that fled, and was frightened at the very flash the mischievous piece gave when it went off), coming nobody knows how or from whence, in a moment puts a period to the brave designs and the life of one that deserved to have survived many years. This considered, I could almost say, I am sorry at heart for having taken upon me this profession of a knight-errant, in so detestable an age; for though no danger daunts me, yet it affects me to think, whether powder and lead may not deprive me of the opportunity of becoming famous, and making myself known throughout the world by the strength of my arm, and dint of my sword.*

The hostility of the ruling caste to guns and gunnery lasted long. As, step by step, gunpowder changed the whole aspect of warfare, the diehards and brass-hats of those days fought their losing battle with a tenacity worthy of better ends. Like most people who detest an innovation, they believed, or affected to believe, that the innovators were immoral or infidel. Even in 1676, an English writer began his dissertation, 'A Light on the Art of Gunnery', as follows: 'The reason, wherefore my first discourse is of Gunners, is only because many times it falleth out, that most men employed for Gunners, are very negligent of the fear of God.'

I should hate to hazard a guess as to the truth of this statement. But it was clearly true, for generations after generations of fighting men, that gunners were thought of as specially terrible fellows, likely to be in league with the devil. And it was also true that the armoured fighting men tried every way they could to suppress or circumvent them. And men clung to their

*Cervantes, *Don Quixote*, Everyman's Edition, vol. 1, pp. 318–19.

armour for centuries after it had lost fighting value. In 1807 the Tsar of all the Russias used against Napoleon's armies, in Poland, 1,500 Bashkirs wearing chain mail and armed with bows and arrows. They were as valueless for fighting as the Lancers of the Viceroy of India are today.

Among the toy soldiers I played with as a child there were some specially proud and haughty horseman called Life Guards. They, and their natural enemies the French Cuirassiers, wore breastplates of steel. It was the thrill of my young life to see these toy soldiers, come alive, very big and clattering most impressively, ride down from Buckingham Palace before the first Great War. But they *were* toy soldiers: when I saw them, in 1912, the great period of the armoured man on horseback was five hundred years away; the last armoured knights of Europe had charged the Turk three hundred years before.

# 5. Castle and Gunpowder

One of the things I have deliberately left out of this book, until this chapter, has been fortification. Throughout the history of warfare, sieges have often been among the most important types of battles, and have often been decisive. Now that we have reached, in our history, the beginnings of gunpowder, it will be as well to go back through the past periods and note some parallels in the development of siegecraft to the tendencies already described.

The walled and defended city has existed since there were cities; in fact the first growth of cities was probably as much for defence as for other social ends. During the first unarmoured period it seems probable that the walled city or little town was the only normal fortification. There were a few less permanent fortifications; the Greeks besieging Troy built barriers of some sort to protect their ships and the camp alongside it. But these 'field fortifications' were rare and would normally only occur in the case of sieges.

Assault on fortifications was always difficult in this period, and was usually undertaken by trick – the Trojan Horse – or by a straight storming of the walls or gate. Battering rams must have been used early in history, usually against gates; but we have no knowledge of other 'machines' until we get to the first armoured period.

Armour permitted the attacking force to get near to the defenders; the defenders were, of course, sheltered by the walls, and armour gave to the attackers a similar though weaker protection. The first development during the period of the armoured foot-soldier that we need to note is the use of tight-packed shields covering a group of men who approached the wall or gate under this cover. Soon this became the 'testudo' or tortoise, a regular military formation for the job. The shields linked together looked like the scales of a tortoise. Much later the same purpose would be served by a pent-house or movable hut, usually made of hides, later still of tiles and hurdles covered with earth; this was used as a protective roof to cover attacking forces from the arrows and burning liquids and heavy stones showered down upon them from the walls.

Protected by their shields, or by such a covering, the attacking forces would bring forward their ram, which at first would be carried and swung simply by men's hands. It might easily be the largest tree of the countryside near the city. Later it was found more efficient to support the weight of the ram on two perpendicular beams and on ropes or chains from these beams; then fewer men could swing the ram farther and with more effect.

Besides the rams there were other 'engines', some of which we shall describe later. These engines were hard to move, and usually easier to construct in their emplacement to defend a city than to build elsewhere for the attack on the city. We have already mentioned the development of these machines about 300 B.C., and the use of some of them by Alexander the Great as a field artillery. As the armoured period developed away from the simple solidity of a homogeneous army, and towards more complexity and co-ordination, there was a considerable increase in the numbers and importance of these machines. In 149 B.C. when Carthage surrendered, the Romans claimed the capture of two thousand catapults from the Carthaginian army and from its stores in the city. This means that there was probably one stone-throwing 'piece of artillery' for every hundred soldiers in the Carthaginian army. This is quite a high proportion of artillery to troops; just to take a comparison, Wellington's army at Waterloo had one gun for rather more than four hundred men.

Carthage probably had this considerable amount of artillery because it was a trading and commercial and mechanical empire; machinery is a natural product of craftsmanship and large-scale production for trade. Syracuse, before Carthage, was also a great commercial city; and it was in Syracuse that artillery seems first to have become important. At the time when Rome destroyed Carthage the Romans seem to have had relatively little artillery, but when Rome replaced her rival as the centre of the world's trade and the monopolist of its most ingenious products, the Romans began to give their soldiers more and more 'machines'. About two centuries after the fall of Carthage, the full equipment of a legion of six thousand men consisted of ten catapults and sixty *carrobalistae*. The latter were light field pieces mounted on wheels, and the legion would take them with it wherever wheels could move. The catapults on the other hand were heavier machines, normally used only for the defence of fortifications, and for the attack upon them.

The first portable fortification of importance in warfare were the stakes carried by the Roman legions. Each legionary on the march carried a long hardened stake of wood, and when the legions went into camp they planted these stakes to make a fence or wall, and also if possible dug a deep ditch outside the wall.

The Roman entrenched camp was a definite part of the Roman tactics. Its wall provided a safe base in which the whole of a Roman force could rest, and from which it could send out detachments or line up for battle as a whole. The surrounding ditches were often fifteen to twenty feet wide; sometimes a second outer ditch would be cut, and a road left between the two along which the Romans could move their troops if they desired. Eventually a linked system of fortifications of this sort, protecting the Roman Empire from German invasion, stretched for 375 miles from Newwied on the Rhine to Ratisbon on the Danube. This *limes Germanicus* was the Maginot Line of its day. It lasted longer than the Maginot Line. It did not last for ever.

When Julius Caesar made a camp on Mont Saint-Pierre, in the Forest of Compiègne, he caused an earthwork to be thrown up twelve feet high with a parapet and ordered that two ditches, fifteen feet deep, be dug in front of it. He built a great number of three-storied towers, linked together with bridges and circular platforms protected by wicker screens, so that the enemy advance was held up by a double fosse and two lines of defenders. The first line was ranged on the upper platforms, whence, better sheltered and overtopping the others, the soldiers hurled their darts greater distances with greater safety. The second line, ranged behind the parapet and closer to the enemy, was protected from their barbs by those who fought on the platform above.

In another case when Caesar drew trench lines to block an army of Gauls into the town of Alesia, he dug special trenches five feet deep and planted the bottom of these trenches with sharp wooden spikes. In front of these he had smaller traps made, about three feet deep, in which the stakes were so hidden that they only protruded for a foot. Briars and other bushes were grown to hide these stakes; the whole effect must have been very like that of a well-kept barbed wire entanglement in modern warfare. His men seem to have shifted about two million cubic metres of earth to make the trenches.

The Roman legion, when properly entrenched, was scarcely ever successfully attacked by its opponents. It had therefore the immense advantage of being able to wait safely within its camp until some diversion had divided the opposing forces, or some addition of strength had reached the Roman commander.

The Romans did not only use their stakes for the camps that were their main military fortifications. They used them also for minor tactical jobs, such as the blocking or narrowing of a ford. If you are foolish enough to go swimming tomorrow in a shallow and muddy patch of the Thames not far from Brentford, at very low water, you may hit your foot on a fire-hardened stake of wood; there are still a few of these stakes in the river bed, put there by Roman soldiers nearly two thousand years ago. This part of the river was once fordable, as the name Brentford tells us. The Roman sentinels sent to guard this ford, as the legions were pushing up through Britain, narrowed it with their stakes so that the enemy could cross at one point only. And when the Roman Empire settled down to fairly permanent frontiers along the *limes Germanicus* the general idea of Roman fortification degenerated from the camp, organized for all-round defence, to the line of stakes or entrenchments spread as a thin linear crust over hundreds of miles of territory.

If you had gone to the Maginot Line in 1939 you would have seen stakes very like these Roman stakes, made out of new materials but arranged fundamentally in the same way. Part of this line was defended by 'iron asparagus', long lines of steel rails, cemented into the earth so that their ends stood up like jagged spikes. These were barriers against tanks. They were not arranged as the Roman camp had been, for all-round defence of areas from which a striking force could be launched; they were arranged as the later fortifications of the decaying empire were laid out – mainly as a thin line. Once that line was pierced its value went.

The Romans built walls across England and southern Scotland, as the Chinese two or three hundred years before them had built a Great Wall to keep out the raiding tribes of the north. But these walls were not at first intended to be garrisoned in such a way that the enemy was always kept beyond them. They were permanent obstacles to any raider; but their main military value was that they prevented a raiding party getting back home with the loot. Burdened with animals and stores the raiders would try to return, and would be caught by detachments from the garrisons on each side of the point where they had broken through. When the raiders became strong enough to become invaders, when Goths or Germans wanted not only to break into the Empire but to stay in it, these walls lost much of their value and significance. But because those who created them came to rely on them more and more, at these same periods a larger proportion of the Roman defending troops were tied to these wall-lines with all the disadvantages of passive defence.

When the Roman legion became a wall-garrison of relative

unimportance it was destroyed; and with it ended the Roman camp. Also, by one of the ironies of history, the classical artillery that had developed so much came to an end at the same time. This artillery had been one of the factors that, to use a philosophical term, 'negated' the Roman legion; it was itself negated by the ending of the Roman legion. It was too clumsy to accompany an army of horse-archers. It was too heavy to be moved by roads when the good Roman roads fell into disrepair. It found no useful targets; cavalry in open order were poor targets and fleeting ones; and there were no longer many important fortifications at which it could hammer. The only fortifications left were the cities, and these became – with the breakdown of the Roman peace – small, poor, scarcely worth robbing.

Then 'Out of the spent and unconsidered earth, The cities rise again.' And first the ram comes back into warfare, and then other engines are invented or remembered for the assault on cities and later on the medieval castles.

The ram was always a clumsy instrument and could normally only batter a hole in a single line of fortifications. If there was a double line, one wall behind the other, the men handling the ram would get in great difficulties as soon as they had breached the first wall. When a city or castle was well defended the outer wall would be lower than the inner wall, and the attackers who seized a portion of it would find themselves still exposed to a fire against which they had no reply. The men trying to use the battering ram against the defenders of the second wall would usually be subjected to a fire from three sides. Their usual recourse, against such systems of walls, was to build a ramp that brought them to the same level as the defenders, or to build towers that they could move up to the walls, in order to fire down on the defenders from the tops of these towers. These had to be made of wood and could be set on fire.

When rams were employed in the siege of Jerusalem in 1099, the defenders used forked beams to push the ram to one side or downwards so that it could not be swung. They also used great pads of sacking which they lowered with ropes from the tops of the walls to act as buffers; the rams hit against these pads, instead of against the parts of the wall that previous blows had weakened.

The development of the ram, and of the bore (a long pole with a sharp iron point, rotated to pierce a wall) made it necessary for fortifications to consist of two or more strong concentric perimeters. Usually the cities in the Dark Ages, and in the early Middle Ages, were too poor to keep up or even attempt to build this solid double fortification around them. Therefore for most military purposes the keep or castle, within the city or not far from it, became of greater importance than the walls of the city itself. Sometimes the castle, as at Lincoln, would be big enough for most of the citizens to take shelter within it if a raiding enemy appeared who might pierce the city walls. And as Feudalism developed, the Feudal nobility built their own castles either where they themselves lived or at strategic points in their territory. These fortifications, organized for all-round defence, can be considered the natural military forms of fortification in the period when a small striking force of armoured knights was the main factor in battle. And like the Roman camps they were, from a military point of view, secure bases from which a striking force could emerge at the right moment for battle.

Here then is a striking parallel between the two armoured periods that we have been considering: in each case the 'high point' in military fortification is a camp or castle organized for all-round defence, and intended to hold out for some period when surrounded by the enemy. It is not a line, but an 'island of resistance'. We shall see this principle of fortification return when we come to the third armoured period, that of the tank.

Two weapons already mentioned, the ram and the bore, were mainly responsible for the doubling of medieval walls. Other weapons forced further development. The siege-tower, with crossbowmen or small catapults upon it that could bring fire to bear against defenders on top of high walls, made it necessary for the second wall to be spaced further away from the first. Ludwig Renn makes this very clear in his book on the relation of war to society: 'As the range of arms throwing a projectile improved the distance between the two walls had to be increased. The defensive walls of Constantinople lie very close together indeed, whereas the distance between the defensive walls of Marienburg in East Prussia, one of the fortresses of a knightly order, is considerably greater, because this fortress was built towards the close of the Middle Ages, when the crossbow, which shot considerably farther than the ordinary bow, was already in use. The distance between the defensive walls of Nuremberg is greater still, because the fortress there was built at a time when firearms were already in use.' And Renn concludes: 'Here we have a principle of fortification which still retains its validity today for defences built in a series of lines: the distance from line to line is determind by the fire range of the attacker.'*

Besides the crossbow, and probably before it, the men of the Dark Ages and the Middle Ages had a much bigger and clumsier machine made in the same shape. The cord was pulled back by small hand winches, and this machine threw bolts or javelins with a fairly flat trajectory. Another type of machine, more like a howitzer than a gun, was the mangon. It was developed from the machine that the ancient Romans called the *onager*, or wild ass, probably because it leapt off the ground with all four feet when released. This consisted of a wooden base from which rise two stout posts, between which were twisted a double or quadruple set of ropes. A beam was then twisted into the ropes in such a way that when one end of it was hauled back, the torsion of the ropes exerted a very considerable force to rotate the beam. A spoon-shaped hole was made at one end of the beam, or a sling attached to it, in which the engineer placed the projectile – a rock or a ball of lead. When the machine was 'fired' by releasing the beam, the ropes, untwisting, would throw the projectile high in the air over the walls that were being attacked.

During the twelfth and thirteenth centuries a slightly more

*Renn, *Warfare, the Relation of War to Society*, p. 119.

accurate machine called the *trébuchet* became the main siege engine. It consisted of a long horizontal pole, balanced on a pivot supported by two uprights placed much nearer the butt end of the pole than the working end. These uprights raised the pole some way from the ground. The working end, the longer part of the pole, was pulled down to the ground and held down by catches; then the missile was placed in the spoon-shaped hole in it or in a sling. The butt end was loaded with heavy weights of iron and stone, fixed in a kind of box or in some cases bound to the pole with cords. When the catches were released these weights pulled the butt end down, the other end of the 'see-saw' flung the projectile high and fairly far.

Trenches were seldom used by besieging parties in these days, but breastworks of twisted hurdle, generally covered with a coating of hide, were used to make covered ways of approach to the 'batteries' or to the enemy walls. And mining and counter-mining played an increasing part in siegecraft; the besieging forces would try to mine under a piece of wall, propping up their workings with timber; then they would set the timber afire and hope that part of the walls would collapse. This method of attack was used by Philip of Macedon when he besieged Constantinople in 340 B.C.; exactly the same method was sometimes used by Crusaders fifteen hundred years later. But mining did not become really important until gunpowder was understood and could be used for it. In 1503, at Naples, Peter of Navarre successfully blew up part of a castle by means of powder. Mining remained a hazardous business. The attackers had to dig deep under moat and castle, and the defenders might flood the hole from the moat or dig through on their own side to fight hand-to-hand in the dark.

The weakest point in any fortification is normally the ordinary point of entry, the gate. The ram was used against a gate when possible; wooden gates of medieval castles could be set on fire or a mine driven under the gate towers. When gunpowder came into use a special form of portable mine for blowing in gates was invented; it was called the petard. It was a heavy metal bowl or pot in the shape of a big top hat, which was filled with powder and either spiked on to the wooden gate of a castle or propped against it with stone or wooden backing. When the fuse had been lit the force of the explosion blew the gates inward – and the petard outward towards the attackers. The unwary engineer who stayed too long or too close, after the fuse had been lit, could be 'hoist with his own petard'.

In those Elizabethan days 'when the world was so new and all', even the high command new and not averse to new weapons, a British force went into action, according to a contemporary ballad:

> With a new shippe and a new Generall
> And a new noble Lord High Admirall
> With a new device to batter an oulde walle
> And a new peatarre to make the gate falle.*

Before we deal further with gunpowder, 'Greek fire' should be mentioned. This has been described as 'wet fire' and is said to have been invented by a Greek named Callinicus in the seventh century. It was mainly composed of sulphur and quicklime, which took fire when exposed to moisture and was projected through a siphon or nozzle. Sometimes arrows were dipped in it before firing, or huge pots containing it were flung on board enemy vessels. The fires caused were extremely difficult to put out. At the siege of Acre in 1190 by the Crusaders, a Damascene engineer burned all the siege-machines of the invaders by flinging jars of the fluid upon them. Earlier on, in the year 717, the great Arab leader Maslama, brother of the Caliph Suleiman, was attacking Constantinople in the campaigns to extend the Empire of Islam. He attempted a land attack, but it proved hopeless in face of the fortifications of the Byzantine engineers. So Maslama entrenched his army, surrounding his camp with a deep ditch, and resolved to compel the surrender of Leo the Isaurian, Emperor of Byzantium, by blockading Constantinople. He instructed his brother Suleiman to send a squadron of ships to force their way through the Bosphorus, and to cut the city off from supplies from the

*Field, *Old Times under Arms*, p. 16.

towns on the Black Sea. This squadron arrived in September 717 and got under way to sail north of the Golden Horn. There lay Leo's fleet in harbour, protected by the great chain suspended between two towers from each side of the harbour's entrance. As the blockading ships came round Scraglio Point they were thrown out of line by the strong current. Immediately Leo ordered the chain to be lowered and his galleys sailed out and poured Greek fire on the enemy's ships, destroying twenty and capturing others. The attack was so well directed and so decisive that it largely contributed to the defeat of the blockade.

Greek fire was an excellent defensive weapon against besiegers, used in reply to the catapulted stones and balls of lead projected over the castle walls by the early siege engines, such as the *balista* and the *trébuchet*. These remained the great weapons of siege artillery until the invention of gunpowder. Then came the period of the bombard of mortar and the gun or cannon.

'Bombards' were used as siege weapons, to throw projectiles in a high trajectory over walls, and cannon as flat-trajectory weapons for direct discharge against walls. Cannon needed large amounts of gunpowder; they also had to have long barrels, so they developed rather slowly. Towards the end of the Middle Ages cannon were used in the field. At the battle of Crécy the English armies used two guns; it is in the records that the army only had twelve gunners. Warring against the Swiss, the Duke of Burgundy used cannon mounted on wagons, but they were not very successful. In the middle of the fifteenth century, eastern armies began to demonstrate the qualities of the new 'pots' or bombards in siege warfare.

In 1453 the Turks under the Sultan Mohammed II – the Conqueror – besieged Constantinople. Mohammed's artillery was formidable, his best bombards being cast for him by a Hungarian cannon-founder named Urbán. These guns threw stone shot thirty inches in diameter and weighing 1,200 to 1,800 lbs. Each 'howitzer' required 60 oxen to drag it, 200 men to march alongside to keep it in position, and 200 more men to level the road. Mohammed had a total of 14 batteries, consisting of 13 great bombards and 56 smaller weapons of all kinds. This artillery greatly assisted the Turkish conquest of Constantinople. And these guns lasted a long time. One of the Turkish guns of 1453 survived till 1807, when the 700-lb ball it projected shattered a mast of Admiral Duckworth's flagship in the Napoleonic wars.

Even heavier projectiles were known. The Venetians in the fourteenth century produced a shell consisting of two hemispheres of stone or bronze, filled with gunpowder, and fired by primitive fuses attached to it, which hissed as they went through the air. These shells sometimes weighed as much as 3,000 pounds according to reports accepted by the historians.

Incendiary shells were also used in the fifteenth century, to set fire to wooden buildings. Sometimes flares were fired in order to light up the enemy lines for the pikemen. Fireballs were also used to blind the enemy. These were solid balls of metal heated in a furnace before being projected. Some of these 'shells' were designed to break up in small pieces when they struck an object. There were, in addition, forms of caseshot, large numbers of small pellets fired straight at the infantry.

While the cannon fulfilled the purpose of the ancient ram, the mortar or bombard was the gunpowder version of the *trébuchet*. The mortar usually had a very short and thick barrel and looked like a pot. The chief advantage of the mortar lay in its relative lightness. And it was nearly always easier to fire effectively over the top of walls, causing damage within, than to destroy them by continual pounding at the outside.

We have already seen how feudalism, the régime based on the power of an armoured cavalry, suffered its first defeats at the hands of the yeomen archers. These men, working for pay, professional soldiers rather than feudal retainers, were the forerunners of new social forces and new classes that were preparing to destroy and remake the feudal world. The ending of feudalism and of armour did not occur rapidly. Cromwell's soldiers were 'Ironsides' and wore a heavy breastplate designed to turn a pistol bullet. At the same period John Sobieski, who became king of Poland by his victories over the Turks, was fighting with a heavily armoured cavalry some of whom wore heavy metal 'wings' designed to make a loud noise as they rode; the noise was intended to frighten the enemy. Some of these Polish knights were strapped to their saddles so that they could not fall off even if wounded: this heavy cavalry was used as a shock force that could break through any infantry line opposed to it.

But long before this, in places less 'out of the way' than Britain or Poland, firearms were beginning to master armour and at the same time to master the castle. The social results of this were first emphasized by Friedrich Engels in his *Anti-Dühring*:

Firearms required industry and money, and both of these were in the hands of the burghers of the towns. From the outset, therefore, firearms were the weapons of the towns, and of the rising monarchy drawing its support from the towns, against the feudal nobility. The stone walls of the noblemen's castles, hitherto unapproachable, fell before the cannon of the burghers, and the bullets of the burghers' arquebuses pierced the armour of the knights. With the armour-clad cavalry of the feudal lords, the feudal lords' supremacy was also broken; with the development of the bourgeoisie, infantry and guns became more and more the decisive types of weapons. (p. 190.)

In the next chapter we deal with the development of the musket and of field artillery. In this chapter on siegecraft we need to note that the use of gunpowder gradually decreased the importance of the castle, but at the same time the growing wealth of towns increased for a period the importance of the strongly fortified town. Cities as big as London or Paris could no longer be strongly defended; they were too large and they were always spreading beyond their walls. But little stable fortress towns, such as Tournay or Badajoz, became of increasing importance and were for a time almost the only form of fortification that mattered much in warfare. The defences of these towns at first developed in the direction of complicated 'out-works'. The fire of the defenders' cannon could not be directed so easily to the foot of the walls from little balconies built out from those walls, as the fire of more primitive weapons had been. It was therefore necessary to push out salients in front of the walls, strong points that were called 'horn-works' or 'redans'. From these the defenders' artillery could fire to the flank along the walls and prevent the attackers' approach to them. But gradually it became impossible to expect ordinary stonework to stand up to the destructive effect of improving artillery. Napoleon Bonaparte, writing a précis of the wars of Julius Caesar, pointed out that 'the arms of the ancient world ... called for upstanding strong points with high towers and walls; modern arms make necessary low forts covered with slopes of earth that mask the masonry.'

With this development we reach the beginning of modern fortifications, to be dealt with later in this book.

# 6. Musket and Bayonet

Few of the weapons that we have been describing have any modern interest to us. Modern mortars and guns have developed directly from the first bombards and cannon; but sword and spear and battleaxe have gone into the discard of time, the arrow has gone, and with these weapons have passed away the tactics and formations of the armies using them. Here at the entry to modern war, war of firearms, infantry, field artillery, it is worth while repeating that some of the patterns of modern war are parallel to those of the past: in new and more complicated forms the same processes work through to similar conclusions. In the modern unarmoured period that we have now to consider the first main process was the slow development of an army that could hit like a single heavy hammer.

We have seen this process before during armoured periods. Fighting mainly by shock weapons, the armoured infantry of classical times or the armoured knights of the Middle Ages first grouped their forces into 'phalanxes', with which they could break through an opposing line. Why does this process come back again in an unarmoured period? In the previous unarmoured period, after the defeat of the legion, there had been no such development. The answer seems to be that the new weapons were too heavy and clumsy to be carried on horseback, and only a few could be pulled on wheels. Therefore the change away from the shock tactics typical of an armoured period was gradual; it did not come fully for three or four hundred years. During those years 'unarmoured shock' developed. The defeat of the armoured knight and the steadily increasing importance of infantry implied a reduction in tactical mobility. In the previous unarmoured period the horse-archer had been tactically mobile; the musketeer was much less mobile. And there was another change that made shock weapons of considerable importance during the time that armour was gradually being wiped out; the musket, which was very slow to fire, was substituted for the bow and arrow which fired rapidly. (The longbow in skilful hands could keep several arrows in the air, one following the other.)

Opposition to the change from bow to musket was powerful. As an example, Colonel Sir John Smyth wrote to the Privy Council in 1591: 'The bow is a simple weapon, firearms are very complicated things which get out of order in many ways ... a very heavy weapon and tires out soldiers on the march. Whereas also a bowman can let off six aimed shots a minute, a musketeer can discharge but one in two minutes.' Many other John Smiths felt the same distrust of the new arms.

Men equipped with the first primitive firearms had only a chance to fire once against an enemy charging them. Defence by longbow could almost be defence by fire alone; when fire-arms were the principal weapons of the defence, they had to be strongly supported by shock weapons for use at close quarters. Otherwise the line of musketeers was likely to be caught while loading and unable to fire.

The principal shock weapon of the period is the pike – a spear used not so much for thrusting, as the ancient spears had been, but more as a portable fortification. The infantry of a medieval army had also contained a number of men with pikes, but when the bow was given up and the new clumsy expensive firearms came in, the proportion between men armed with projectile weapons and men armed with these shock weapons altered. The firearms were at first few; the pikes were many. For these reasons the opening part of the unarmoured period of modern war shows some of the charac-teristics of an armoured period: importance of shock weapons and the development of drilled, homogeneous 'heavy' forces, to which light forces were auxiliaries. This is in briefest

summary the period from the first firearms up to the army of
Frederick the Great.

The first firearms were, as I have stated, greatly disliked by
the foremost soldiers of the day. An entirely unambiguous
attitude towards them was taken by the Chevalier Bayard,
whose name is still a symbol for noble soldiering; of Bayard
the phrase was first coined 'without fear and without re-
proach'. He considered it correct and Christian to hack at a
man with a sword or run him through with a spear; these were
things that had always been done. But it was devilish to shoot
him from a distance; it was unfair that the anonymous churl
with an iron tube and some gunpowder, and a great slug of
lead, could abolish a knight before the knight realized what
was happening. Therefore the Chevalier when he took prisoner
any man who, by his dress or equipment, could be seen to have
carried a firearm, hanged his prisoner on the spot.

The first firearm was a short metal tube with a straight stick
roughly fixed to it. The stick was held under the arm when
firing, and the weapon was, as we should say now, 'fired from
the hip'. Actually it was fired with the stick between the armpit
and the hip. This hand-gun, later arquebus, had a small hole in
the side of the barrel into which a pinch of gunpowder was
shaken after the weapon had been loaded from the muzzle end.
Loading was a business; the soldier had to hold his weapon
vertical and pour into it a measured quantity of rough-grained
powder. Then he took a thick wad and rammed it down on top
of the powder. Then he dropped down the slug or ball, which
was usually so roughly shaped that it would jam and have to
be forced down with a ramrod. It was often necessary to put
another wad on top of this, if the projectile was at all loose in
the barrel; if this second wad was not put in, the projectile
might, when the weapon was being moved or levelled, slide
down the barrel or even fall out of it.

Powder was usually carried in a horn, and the weight of
charge had to be guessed. There were no neat little cartridges,
each containing the correct weight of propellant. There was
therefore very little uniformity about the range of the weapon;

and powder might be damp and not give much more than a fizzle. The muzzle velocity was low, even if the powder was dry and shot correctly shaped and sized, and so the shot had to be a large one. A light bullet travelling slowly would not have gone through armour. For this reason the calibre of the old arquebus, and of the musket that followed it, was far greater than that of the modern rifle; it was more like that of a 12-bore shotgun.

The arquebus could usually propel its heavy projectile between 200 and 300 yards. But this was extreme range, and normally it would be used at less than sixty yards. At twenty yards a skilled man could hit a small haystack, or a group of four men riding abreast. Then it was discovered that if the barrel was made longer and the charge of powder increased and the projectile moulded more carefully, fire could be a little more accurate at longer ranges. And the Spaniards introduced the first real musket, about the middle of the sixteenth century. It fired a lead ball weighing about one and a half ounces.

The barrel of the musket was of much heavier gauge than that of the arquebus – and also longer, in order to use the full force of the propellant, which burned fairly slowly – so that the weapon was at first too heavy for a man to fire it without some support. Each musketeer had an assistant who carried his weapon before action, and during action helped him to prop it on a specially made pole, one end of which was stuck in the ground. Experiments were also made with a two-legged pole, like the bipod mounting of a modern light machine-gun. It was found inefficient to balance the weapon on a pole which was connected to it at its centre of gravity; this allowed the weapon to slew about too much. These props were therefore normally at the far end of the barrel and the musketeer took half the weight of his 'engine'. It soon became clear that it was easier for him to have the powder pan, which he had to light, high up rather than low down; therefore he must put his weapon to his shoulder rather than tuck one end of it under his arm. So early muskets had the first roughly shaped stocks. But the weapon was still so inaccurate that there was not much point in looking along it, for aiming, and therefore the stock was not so shaped as to bring the barrel up to the level of a man's eye.

It will be remembered that the full development of the bow and arrow came only with the longbow, which was so tall that the man wielding it had to pull the string back to his eye or ear and therefore could look along the arrow for aim. In the same way the full development of the musket only came when the weapon had been used for a long time, had been accepted by the nobility and gentry, and was made in particularly fine and handwrought specimens for sport, particularly for shooting birds. Against a moving target the sportsman needed, not sights, but stocks rightly shaped – even the best shotguns of today have scarcely anything you can call a sight on them, and those most skilled in their handling seldom try to use the sights. But they do have the barrel so placed that the eye naturally goes along it towards the target. And gradually through centuries this development, implying a certain shape and angle of the stock, occurred in the musket.

A development also occurred in the means of firing. At first it was necessary to have a burning 'match' with which to light the little pinch of powder beside the touch-hole. The act of applying this match upset any attempt to aim. And it was inconvenient to carry a smouldering piece of material about with you; it would always go out at the wrong moment, in any shower of rain. If carried in the cap, as it usually was, it might easily set cap and hair on fire. It had to be blown on to get it burning well before it was used; and this might take one or two puffs or might take some time. In spite of all these disadvantages these muskets lit by matches were in use for a considerable time. They developed into the match-lock, which had a little piece of metal at the side of the weapon to hold the burning match (made out of cotton soaked in saltpetre), and when a trigger was pulled the burning match was brought into contact with the gunpowder. This permitted more accuracy in aim. Later in Germany and Spain there appeared the wheel-lock, with a metal wheel at the side of the priming pan. A piece of iron pyrites pressed against this wheel, which had cogs on it. The trigger rotated the wheel, or released a spring to rotate it; the wheel struck sparks from the pyrites. This was too complicated a mechanism to replace the match-lock for ordinary infantry, and the match-lock did not disappear until the flint-lock appeared. This is the familiar thing seen in many old sporting weapons: a flint is held above the pan; the action of the trigger and a spring brings this rapidly down against a cover of metal that protects the priming of powder from damp. As the flint pushes this movable steel pan-cover away, the priming powder is exposed to the sparks falling from it.

Gradually during the seventeenth century the flint-locks replaced other types of musket. Percussion caps, which would go off when struck by the 'hammer' of a musket, were invented early in the nineteenth century, but were not used by the British army for many years after their invention. After a prolonged trial by the army authorities they were adopted in 1835. By 1850 British conservatism could no longer resist the development of the rifle, though many of our troops in the Crimean War (1853–6) were still armed with the percussion musket 'Brown Bess'. No more muskets were made for issue to the British army after 1855.

The principle of rifling a weapon had been discovered early in the sixteenth century.

The first muskets were more accurate than the arquebus, but not accurate enough. Within the smooth-bored barrel the projectile hopped about, as the gases produced by burning the propellant blew past it now on one side and now on the other. The direction it took when leaving the muzzle was somewhat determined by the direction in which it was hopping at that moment. The lead ball, in later muskets, was covered with a heavy grease or wax, or a plaster, so that it would not bounce around so much; but rifling gave a better and more mechanical positioning of the projectile in grooves cut within the barrel. And a projectile from a rifle flies more accurately because it is made to spin, and like a spinning top keeps itself in the same line.

A rifled musket was even slower to load and fire than an ordinary musket. The bullet had to fit tightly in the grooves of the rifling and it could not be pushed down from the muzzle without a great deal of effort. Ramrods had to be heavier and stouter, and often the bullet had to be forced down by the use of quite a heavy hammer on the end of the ramrod. Therefore the first rifled muskets were developed for sport, and for hunting. The rifle did not come into war until men who kept themselves by hunting, George Washington's Colonials, used it for their sniping. But that is a later story.

The early unrifled muskets could be dangerous at 400 yards distance, but a man was most unlucky to be hit at far shorter ranges. They were too inaccurate for the musketeer to be sure of hitting a single man moving at twenty yards range. It may seem surprising, when this is considered, that armies remained in the close order natural to them when they carry armour and use shock weapons. Why did not the soldiers of Tudor or Cromwellian times scatter into small units, or even into individual skirmishers, so that they could not be hit by the fire of their opponents – except by accident? The answer to this lies in the fact already emphasized: the slowness of fire of the musket made it necessary for the men with muskets to be protected by men with pikes. No other infantry weapon could hold up a cavalry charge; and the pike could only do this if the pikemen were massed closely together in several ranks.

The complicated motions of loading a musket – about sixty motions were necessary, and at one period over thirty orders – could be carried out most rapidly if men had practised these motions until they were thoroughly accustomed to them. Therefore drill and the drill-sergeant became of great military importance. Men were of course drilled in the actual movements they would carry out in battle; but the drill was more difficult than the keeping in step and in line that was the essence of the drill in Greek or Roman days. The musketeer must not only do his job neatly and quickly; he must also do it in a confined space with the pikes on each side of him. And the pikemen must be drilled also so that they would always present a hedge of metal points from which any cavalry would flinch.

These musketeers, the primitive infantry of our day, were men using a mechanism in battle, and gradually they were drilled and trained to become themselves machines. Victory lay in their imperturbable march forward under fire until they were at so close a range that their clumsy weapons could cause disorder in the enemy line. Safety under the enemy attack lay in their holding their fire till the moment when enemy cavalry charging were only a few yards from them, and then as one man firing a volley that would break the enemy charge. During all this period when shock was still of great importance, the best infantry became more and more machine-like, until this man-machinery had been perfected by the greatest of the Prussians, Frederick the Drill-master.

Cavalry meanwhile, for some time, came to rely less on shock than on a combination of shock and fire. The horse-pistol, the sort of thing Dick Turpin carried, was one of the main weapons used by Cromwell's cavalry, for example. Cavalry could with some safety approach fairly close to an infantry formation, which would naturally hold its fire because it feared that it might be reloading when the cavalry charged. Then horsemen would ride up almost to the pike-points and discharge their clumsy pistols in the faces of the men on foot. The horsemen hoped to put their opponents in disarray or cause some of them to turn tail, when they could be ridden down with swords. Even when Cromwell's cavalry attacked other cavalry, their aim was first to discharge their pistols effectively and then to use their weight and solidity for the shock of the charge. Discipline was therefore very necessary for the infantry, who without it would have been exposed to cavalry attack. And soon it became useful or essential for a division of function to take place between various ranks or sub-units of the musketeers, so that one portion could fire their volley, and another portion remain with muskets ready to fire if a cavalry charge occurred. In the sixteenth century two methods of volley firing for musketeers were worked out. No problem had arisen when the musketeers were so few that all of them could be in the front rank with a pike on each side. But this gave too little fire, and as more firearms were made the musketeers became proportionately more numerous. The first method of volley firing was used for troops advancing. The front rank, as it came within a short distance of the enemy, would fire and then stand fast, allowing the second rank to advance through it and fire a second volley. Soon musketeers were in three ranks, and a third rank would march through the first two to dress, to blow up their matches, to go through all the rigmarole of firing. Meanwhile the first rank would be reloading, and after a time would be ready to press further forward and fire another volley.

This tied infantry to a slower rate of advance than had ever been the case before. And the second method, similar to the first but in the reverse direction, adopted for troops on the defensive, was unsatisfactory because the first rank would retire after firing. Such a movement to the rear could become a retreat. It was difficult to have a sort of barn dance going on, with the first rank falling back and the second rank simultaneously advancing to occupy exactly the ground they had held. So various other movements were attempted. At the battle of the Boyne, in 1690, the front rank, after firing, fell on their faces, allowing the second rank to fire over them. The second rank then knelt down to permit the third to fire. But there was a snag in this: all three ranks had to reload at once, and therefore the system did not work on the defensive. The only time when it worked was when the infantry were going to charge immediately after they had fired their volleys. Even then it was clumsy, for men who were going to charge were lying down or kneeling.

To have spread out all the musketeers in one line would have made them too thin-spread to stand attack by cavalry or by a massed infantry. They could not fire over each other's heads as archers could 'in the ancient and vulgar manner of discipline', two centuries or more before. Therefore the first sub-division crept into the drilled homogeneous army;

and this sub-division came in the form of volley firing by platoons.

In this method of firing each battalion was divided into two or four platoons or 'grand divisions', as they were sometimes called, and the first rank of each of these sub-units fired at the word of command, then usually these ranks fell back to reload immediately behind the second rank, which also fired. The whole process when fully developed at the end of the sixteenth century went according to this account:

Let us suppose our Battalion drawn up with the Army on the Field of Battle, three deep, their bayonets fixed on their muzzles, the Grenadiers divided on the Flanks, the Officers ranged in the Front; and the Colonel, or in his absence, the Lieut.-Colonel (who I suppose fights the Battalion) on Foot with his Sword drawn in his Hand, about 8 or 10 paces in their Front, opposite the Centre, with an expert Drum by him. He should appear with a chearful Countenance, never in a Hurry, or by any Means ruffled; and to deliver his Orders with great Calmness and Presence of Mind.

The first Thing the Colonel should do, is to order the Major and Adjutant to divide the Battalion into 4 grand Divisions.

As the Commanding Officer will be exposed to the Fire of his own Men, as well as that of the Enemy he is to take special Care that he keep opposite the 2 Centre Platoons while the other Parts of the Battalion keep firing; and he must also take as great Care that when it comes to the Turn of the Centre Platoons to fire, that both he and the Drum step aside and return as soon as they have done, otherwise they must fall by their own Fire.

On the word 'March' the Officers move to the Rear of the Intervals. The Senior Captain posts himself in the Centre 8 or 10 paces from the Rear Rank, the other Officers 4 paces from the Rear Rank, dividing the Space. The Ensigns with the Colours in the Centre Rank on the Right and Left of the 2 Centre Platoons; Sergeants on the Flanks and in the Rear between the Officers. The Drums divided in 3 Parts on the Right and Left in rear of the 2 centre Platoons, all dressing with the Sergeants. The Major and Adjutant on the Flanks.

The Colonel having thus spoke cheerfully to the Men he then gives the word 'March'; at which Time the Drum beats to the March and when the Battalion has got within 4 or 5 Paces of him, he turn towards the Enemy and marches slowly down till he finds them begin to fire upon him; upon which he orders his Drum to cease beating and turning to the Battalion, gives the word 'Halt!' and then orders his Drum to beat a 'Preparative', on which the 6 Platoons of the First Firing make ready, etc., etc.

The Officers and Sergeants of these Platoons are to take great Care that the Soldiers level well their Arms so that their Fire may have Effect on the Enemy and also caution them to wait the next Signal of Drum. (Here the Men ought in training them to be us'd to that of recovering their Arms somethings instead of firing, which will make them take in waiting for Orders to fire.)

The Platoons being presented, the Colonel orders the Drum to beat a second 'Flam' on which they fire and immediately recover their Arms, fall back and re-load as fast as they can, &c., &c.*

At the Battle of Dettingen, in 1743, when the British and Hanoverians won a victory over the French, the British infantrymen, using improved muskets, stood facing the French who were using their artillery and were preparing for a charge. The musketeers advanced to within sixty paces of the enemy and then let them have it. The effect of their fire is described as follows in the diary of an officer of the Royal Welch Fusiliers:

Our Army gave such shouts before we were engaged, when we were about one hundred paces apart before the action began, that we hear by deserters it brought a pannick amongst them. We attacked the Regiment of Navarre, one of their prime regiments. Our people imitated their predecessors in the last war gloriously, marching in close order, as firm as a wall, and did not fire till we came within sixty paces, and still kept advancing; for, when the smoak blew off a little, instead of being amongst their living we found the dead in heaps by us; and the second fire turn'd them to the right about, and upon a long trot. We engaged two other regiments afterwards, one after the other, who stood but one fire each; and their Blue French Foot Guards made the best of their way without firing a shot. . . .

Our Regiment sustained little loss, tho' much engaged; and indeed our whole army gives us great honour. . . . What preserved us was our keeping close order, and advancing near the enemy ere we fir'd. Several that popp'd at one hundred paces lost more of their men, and

*Kane's Campaigns, 1689–1712, quoted in Field, Old Times under Arms, pp. 91–3.

did less execution for the French will stand fire at a distance, tho' tis plain they cannot look men in the face.*

It will be seen that the musket was still extremely inaccurate, but could cause a serious effect if fired at a solid mass of enemy troops at a range of less than a hundred yards. By this time almost all the men in the ranks could carry muskets; pikes were no longer necessary. It was still necessary to have a barrier of 'cold steel' against the enemy cavalry charging at a moment when too many of the musketeers were reloading their pieces – or more rarely against an enemy infantry charge. But this barrier could now be put up by the musketeers themselves; the bayonet had been invented.

Both pike and musket were heavy and clumsy weapons, and it was not possible for the same man to handle both of them. A musket usually weighed 15 lb. or more. In 1647 a gentleman named Puysegur, who came from the town of Bayonne in south-west France, was in command of French troops holding Ypres. As seems to be the custom in those parts, he was short of men; he was particularly short of reliable pikemen. And he could get no reserves.

He himself probably carried one of the short daggers with rounded handles which were at that time manufactured in Bayonne. He must have tried one of these daggers jammed into the muzzle of a broken musket, and found that it fitted and made a tolerable substitute for a pike. He sent to Bayonne for a consignment, and when they came, distributed them among his musketeers, who were taught to plug them into the muzzles of their weapons after the weapons had been fired. The daggers were known as 'Bayonnettes'. By 1663, or perhaps a little earlier, English soldiers were using plug-bayonets of this kind.

By 1671 plug-bayonets had been issued to all French fusilier regiments. It was difficult to rely on these bayonets, because they shook loose from the barrels into which they were plugged the first time men lunged with them. Therefore the plug-bayonet was improved in 1687 by a small socket being cut in

*Field, *Echoes of Old Wars* (1934).

the wood of the handle, into which a wedge or plug fitted to keep the whole thing in place. This change, introduced by Vauban, was at once adopted by the French army, and by 1703 the pike had practically been abandoned in France.

It was impossible to fire the musket while a bayonet of the plug or socket type was inserted in its muzzle. In 1689 the troops of William and Mary of England were defeated by Highland forces under Dundee at Killiekrankie, because these troops had their bayonets pluged in too soon when they should have been firing, and then took them out to fire or reload just before the Highlanders charged. The Highlanders rushed them before they could get the bayonet back into their weapons. It was quite natural that the commander of the defeated force should think up another type of bayonet.

This was the ring-bayonet, which was clipped on to the outside of the musket barrel by a large metal ring. This permitted the musket to be loaded and fired while the bayonet was attached. Infantry could now deliver their fire and charge at once. The pike was no longer of use. But as usual, custom retained the old weapon long after its use had gone; far into the eighteenth century the English army still retained a 'colour guard' of fourteen pikes for every company. And the pike reappeared, to our dismay, in 1941.

Ring-bayonets were in general use for about a hundred years after their invention. Their defect was that they were difficult to fit and became loose if the metal ring stretched in any way. In 1805, Sir John Moore – who at that period was revolutionizing the tactics of the British army – introduced a bayonet that could be fitted rapidly and securely by means of a spring clip – the method in use to this day.

While the fire-power of muskets was increasing during the seventeenth century, because of gradual improvement in the weapons themselves, because of better drill, and because of the invention of the bayonet and the disappearance of the pike, another form of fire-power was at the same time becoming increasingly important. The field gun was becoming more handy and more accurate. It was no longer a cumbrous thing mainly intended for siege purposes but occasionally used in mobile warfare; a differentiation, or 'division of labour', was occurring in artillery, so that armies began to have a 'siege train' of heavy guns that were only hauled up from the base for sieges, and a 'field train' of lighter pieces for normal fighting.

Gustavus Adolphus, King of Sweden, won victories that gave Protestantism the certainty of survival in Northern Europe, early in the seventeenth century, mainly by skilful use of light field pieces as auxiliary weapons. His great aim was always to keep down the weight of his guns, and one of his favourite weapons was a one and a half pounder. Guns began to be far more accurately and carefully made; at first some had been made like barrels out of metal 'staves' bound together with rings that were shrunk on to them; later all were made, or almost all, by casting the gun as a solid lump of metal and then boring down the centre of it to make the barrel and chamber. Gustavus Adolphus had some pieces made of leather strongly lined with metal, but even with the poor powder and light charges of those days such weapons can scarcely have been good for more than a few rounds.

The first mobile field pieces were mounted on wagons. They could only be aimed laterally by hauling the wagon round, and they were elevated for range in even more slow and primitive ways. Some had wedges that had to be hammered in on each side of the muzzle to raise the front of the gun. Others had an apparatus rather like the bars used by athletes for a high jump. The crossbar could be raised or lowered on pegs or notches in the upright bars; and the muzzle of the gun rested on the crossbar and was raised or lowered with it. There was no recoil mechanism for these guns and no springs on the wagons; so the whole thing used to shake itself to pieces after a number of discharges.

Later field pieces were made with their own wheels, and a trail which lay on the ground behind them. The recoil was taken by this trail shifting back along the ground. It was easy to lift the trail and shift it sideways to give lateral aim; but elevation was still given by a clumsy business of wedges until a rough screw arrangement, each turn of the screw filed out by hand, was developed to position the gun.

But the number of field guns, in the sixteenth, seventeenth, and early eighteenth centuries was still fairly small in proportion to the size of the armies. It will be remembered that the Carthaginians, fighting defensive position warfare for their city, had used one 'engine' or piece of artillery for every hundred soldiers. At Blenheim in 1704 Marlborough had one piece of field artillery for 900 men. Napoleon at Borodino in 1812 had one gun for every 240 men. In this great battle at the gates of Moscow there seems to have been the heaviest concentration of fire-power known in the Napoleonic era; the Russians had even more guns than Napoleon, and an even higher proportion between guns and manpower. At the battle of Waterloo, as we have already noted, Wellington had about one gun for every 400 men; Napoleon was better armed, with one gun for every 300.

But with the Napoleonic era we have entered a new phase of warfare, in which artillery becomes more important than it had been in the previous three hundred years. In those three hundred years guns were only an auxiliary weapon, of increasing importance but seldom of decisive effect.

Another weapon or arm was developed during this period to do the same job as the gun. This was the 'grenade', so called because it was compared to a pomegranate. As muskets were inaccurate, and attacking troops had to get within a few yards before they delivered their fire and prepared to charge, it was possible for men to run forward and throw a shell with the fuse burning into the ranks of the enemy. It was a risky job and hard work; the tallest and strongest men were chosen to be grenadiers. Soon there were grenadier companies in each regiment; the use of special units for throwing or bowling these grenades dates from about 1660. By about 1740 the Russian army had separate grenadier battalions; but it is hard to determine whether these men were all in fact armed with grenades. The word grenadier was already becoming a title of honour. Napoleon organized whole brigades and divisions of grenadiers; but these were armed like ordinary infantry. Because the tallest and strongest men had been chosen originally to be grenadiers, and because they had been placed in the position of honour 'on the right of the line', the word grenadier soon became a mere title. Our own Grenadier Guards, formerly the Foot Guards, received their present title after 1815 as an honour for their gallantry at Waterloo. They did not, in becoming Grenadiers, alter their arms or tactics.

Grenadier companies, some of whom were armed with grenades and some of whom had ordinary infantry equipment but were the 'storm troops' of their units, existed in the British army until 1858. But when the rifle was replacing the musket it was no longer possible for whole companies of grenadiers to run up to their enemy's lines and roll, bowl or pitch their grenades among the legs of their opponents. Riflemen (or even men with good accurate muskets) could pick them off before they got within range. Skirmishers prevented their getting near enough to the main lines of their enemies. The hand-grenade therefore began to go out of use during the Napoleonic period, and almost disappeared from warfare for a century, until it was revived by the trench warfare of 1914.

The first phase of modern war, the phase with which we have been dealing, reached its full development with the army of Frederick the Great of Prussia.

Cavalry, by the time of Frederick the Great, had become of relative unimportance in battle. In many campaigns, including some of Frederick's, it was called the decisive arm because its shock tactics, or a combination of shock and fire, actually won the battle after infantry fighting had weakened or loosened the organization of one of the contending armies. But this infantry fighting was in fact decisive: the cavalry only clinched the argument. In a few campaigns, such as Cromwell's, cavalry had been without doubt the decisive arm. By the time of Marlborough, cavalry no longer had great value when employing shock and fire; its pistol fire-power was too little, and the fire-power of the infantry armed with musket and bayonet was too great. A cavalry that checked to fire exposed itself to a destructive volley from a section of the infantry it was attacking. Marlborough allowed his cavalry 'but three charges of powder and ball to each man for a campaign, and that only for guarding their horses when at grass, and not to be made use of in action.'* But cavalry remained of great value for purely shock action, and after Marlborough had pinned an enemy to his ground by powerful infantry attacks, he launched his cavalry against them to cut their formations to pieces.

Cavalry are large targets; it was the increasing fire-power of the infantry that by Frederick's day had reduced the horse to a subsidiary role – that of pursuit. Yet partly through custom, and partly through the need for shock action by the infantry to complete the work that fire-power began, infantry formations remained extremely tight-packed, and Frederick drilled the whole infantry of his army to move almost as one man.

In the three centuries before Frederick there were many developments of warfare that we need scarcely deal with here: the trench systems developed in Northern France, the great fortifications of Vauban, the extreme slowness of sieges, the custom of going into winter quarters and of only fighting one campaign a year. From these and other elements the autocratic monarchies of Europe built up a 'system of war' that Frederick both exemplified and helped to destroy.

The 'system' of slow eighteenth-century warfare derived directly from the political aims of the antagonists and from the social relations between the States and the soldiers they owned. There were few wars in this period that had a national character: war was a continuation of autocratic diplomacy which tried to win a province or a town, at most the succession to a throne, from some opposing but similar autocracy. And the monarchs contending had a very sharp and personal feeling about their own property; willing to grab a town owned by somebody else, they were extremely anxious not to lose any town of their own. They therefore spread their armies out in detachments over wide fronts, seldom and slowly concentrated

*Kane's Campaigns, quoted by Fuller, Decisive Battles, p. 396.

for battle, avoided battle whenever possible, and hoped by siege and manoeuvre against an enemy's communications to secure their limited aims at a very limited cost.

The soldiers of the period were usually conscripted serfs or peasants with remarkably little interest in the wars they were fighting. They had to be kept in camps or barracks for fear they would desert; they could not be trusted to forage for food or other supplies in an enemy countryside. All their supplies had therefore to be brought to them from depots in the rear. These depots could not be very far away; roads were very bad, wagons very slow and clumsy. If the depot was a long way away a wagon would start out with forage for the horses of the cavalry; but before it got near to the army the mules or horses dragging the wagon would have eaten all the forage carried in it. When an army moved far it had therefore to establish behind it a number of depots, and its supplies came stage by stage from one depot to another until they reached the fighting forces.

This question of supply has often influenced strategy and tactics to a very great extent. It still does so. But there probably has never been a system of warfare more subject to the overruling mastery of supply than that of the eighteenth century – at least in Europe. In Asia armies had often been cumbered not only with immense supplies but also with hundreds of thousands of camp followers. A French writer* describes the army of a Mogul Emperor as follows:

The cavalry forms the principal section, the infantry is not so big as is generally rumoured, unless all the servants and people from the bazaars or markets who follow the army are confused with the real fighting force: for in that case I could well believe that they would be right in putting the number of men in the army accompanying the king alone at 200,000 or 300,000 and sometimes even more – when for example it is certain that he will be a long time absent from the principle town. And this will not appear so very astonishing to one who knows the strange encumbrance of tents, kitchens,

*François Bernier (1625–88), in *Voyages Contenant la Description des États du Grand Mogol*.

clothes, furniture, and quite frequently even of women, and consequently also the elephants, camels, oxen, horses, porters, foragers, provision sellers, merchants of all kinds and servitors which these armies carry in their wake; or to one who understands the particular state and government of the country, in the kingdom, namely that the king is the sole and only proprietor of all the land in the kingdom, from which it follows by a certain necessary consequence that the whole of a capital city like Delhi or Agra lives almost entirely on the army and is therefore obliged to follow the king if he takes to the field for any length of time. For these towns are and cannot be anything like Paris, being properly speaking nothing but military camps, a little better and more conveniently situated than in the open country.

In the eighteenth century European armies can seldom have taken with them quite so large an 'organization of supplies'. But in his *History of the Thirty Years War*, Guideby mentions an army of 38,000 fighting men, which was followed by 127,000 women, children, sutlers, cooks and other camp followers. And one of the French armies which fought against Frederick the Great left behind when retreating its officers' 'pommards, perfumes, powdering- and dressing-gowns, bagwigs, umbrellas, parrots; while a host of whining lacquays, cooks, friseurs, players and prostitutes were chased from the town to follow their pampered masters.'*

It will be seen that an eighteenth-century army whose communications or depots were threatened had much to lose. And therefore for a period war became a slow stalemate of manoeuvre in which eight miles was a good day's march, and three towns taken comprised the achievements of a campaign.

Frederick the Great, warring against powerful empires with forces always inferior to those of his opponents, remained to some extent within the rules of this game; but in other and vital directions he had to break up those rules. War was slow, and war was a matter of shock finally, though fire did much to make a decision by shock possible. Frederick determined from

*Campbell, *Frederick the Great, His Court and Times*.

the beginning to alter these ways of war. He would produce an army that could move rapidly. He would produce an infantry that could fire rapidly. But he did so by carrying to its logical limit the drill of his day. His men were drilled to march at a quick-step; they were drilled to fire more rapidly than their opponents. Frederick boasted that they could fire three times as fast. He taught his army to deploy with great rapidity, from a column marching along a road to a single line heading to the flank. The armies he fought against deployed very slowly, and the incredible drill books of the day almost tied them to the ground they stood on. Each unit could perhaps advance or retire; but only with the utmost difficulty and a great expense of time could it face or move obliquely at an angle to its original deployment. Frederick trained his men to face and move in an oblique formation. He also trained them to use their fire-power to the utmost in thin wide-stretching linear formations.

By his greater mobility Frederick was able, on the field of battle itself, to march a wing of his army or even almost his whole army round to the opponent's flank, and then attack it obliquely so that the fire of his men concentrated on and crumpled up point after point along the enemy line. Major-General Fuller writes of him in *Decisive Battles*:

In his earlier campaigns he relied more on the bayonet than the bullet, but soon discovered his mistake; for in his later battles he did his utmost to develop the power of both his muskets and his cannon. He was the creator of the first true horse-artillery ever formed, a weapon so little thought of that, from 1759 onwards for thirty years, the Prussian was the only horse-artillery in Europe. Further still, he was a great believer in the howitzer, because the Austrians, acting normally on the defensive, were prone to hold their reserves behind the ridges occupied by their firing lines. Yet, though so clear-sighted as an artillerist, he never grasped the full value of a trained light infantry, and this is all the more astonishing because at the battle of Kolin the Austian Croats and Pandours were largely responsible for his defeat.*

* *Decisive Battles*, p. 431.

By 'horse-artillery' is meant artillery which is not slowly and clumsily dragged by long teams of horses to the battlefield, and unable to move during action from the point where it is deployed, but artillery that can be limbered up and moved quickly, as quickly as the cavalry it accompanies, to a new point as an action develops. Here again Frederick achieved tactical mobility by patient drill and continuous training of his men. For nearly two centuries battle had almost always consisted of a head-on encounter by two armies meeting face to face. Even Marlborough's battles had essentially that character. Frederick brought back again the tactical manoeuvre, the swing to envelopment. And it is symbolic that at one of his most striking victories, at Leuthen, his flank attack jammed the enemy infantry so closely together that they were, at the vital point of the battlefield, thirty to a hundred ranks deep. His own infantry, two or three ranks deep, could use their weapons to the full; this packed mass of Austrians could use only a small proportion of their muskets. Frederick's artillery could not fail to find a target amongst them. And when towards the end of the battle Frederick's cavalry drove the Austrian cavalry off the field, and attacked the rear of the disordered Austrian infantry, a battle that almost ranks with Cannae had been won.

The army of Frederick the Great therefore stands at the end of a long period of development in which the main process has been the creation of a force capable of hitting, increasingly by fire-power, like a single heavy hammer. But like most of the great developments of history that sum up a period and carry that period to its highest level, Frederick's army had within it the seeds of a new period. He bought mobility with the old currency of drill; but he bought so much of it that his troops in effect entered the new and revolutionary period. when not drill but *élan* would wrench warfare into speed and mobility. Frederick is both the summing up of the age of the drilled musketeer and at the same time, with his horse-artillery, the precursor of Napoleon the Gunner.

# 7. Napoleon

Napoleon was a gunner who became emperor. Previous captains whose mark on the history of war, and of mankind, can bear mention on the same page with his had been other sorts of soldiers. Gustavus Adolphus was a king who liked guns; that is very different from being a gunner who likes making kings. The period that we enter with Napoleon is one in which first the marksman-skirmisher and then the gun, the long-range weapon, uncoffin warfare from the winding sheets of the past, restore movement to it, and decision. The gun and the musket that could be aimed so revolutionize war that the process of change takes on its own momentum, and goes far beyond Napoleon's own innovations, even beyond his understanding.

The political and economic processes that produced the French Revolution, and made possible Napoleon as its heir and as its executioner, are not our concern in these pages. But everyone knows that the Industrial Revolution was producing, during the years before the French Revolution, an age of iron, of simple machinery, of mechanical power. This age opens with the American War of Independence, the formation of the State that was later, in our own days, to carry mechanical power to its highest level of development. And the weapons and tactics of the Napoleonic period first begin to show their shape, symbolically enough, at the place now called Pittsburg. English troops, with as their auxiliaries Colonial Militiamen – of whom a George Washington was the officer – there met and were massacred by French troops whose auxiliaries were Red Indians. The Indian way of fighting was entirely primitive; it would not have seemed out of place to some of the soldiers of the army of the Great King invading Greece from Persia in

480 B.C. They fought by arrow and by ambush; they knew nothing of drill but much of cover. The best-trained troops were useless against them if those troops knew nothing beyond the conventionalized formalities of a European battlefield. In the clash between drilled regiments and the tribes of 'Braves', war returned – as arts must sometimes do – to the primitive; conventions built up by centuries were broken down, so that a new convention embodying the powers of new instruments, and new powers of old instruments, could emerge.

The British army needed to produce soldiers who could meet the Indian allies of the French on equal terms; this need led to the formation of our Royal American Rifles, the first modern infantry. These forerunners of the Rifle brigade wore a uniform designed to hide the man wearing it. All previous uniforms, from the liveries of royal guards or noblemen's retinues, through the red coats of Cromwell's troops, to the elaborate and desperately uncomfortable kits of George III's infantry, had been designed largely to make the wearers obvious. In battles that were like large and brutal games in the open fields, commanders and men needed to know who was on their own side and who was against them. The uniforms they wore were therefore brightly coloured, like the jerseys worn by football teams. They were also often elaborate, even decorative, partly because it was considered good for drilling men into automata to make them slave at polishing buttons and other gear; partly because the richness of a uniform showed the wealth and therefore the fighting resources of the autocrat at whose service was the man within the uniform. Uniform of this sort was a hopeless handicap in the forests of America. The Rifles therefore wore green jackets. And they wore black buttons, as their inheritors do to this day.

From the disaster at Pittsburg (then Fort Duquesne) George Washington was the only officer on the English side to escape alive. Some years later he began to teach himself and the British army, on a larger scale than could derive from one scrambled ambush, the possibilities of the new tactics. These possibilities largely arose from the fact that an accurate long-range weapon now existed. The musket, and in particular the rifled musket, had ceased to be a weapon valuable only at a hundred yards or so against a mass of troops. It could be used against a single man, even against a man partially under cover. It could be used at ranges so great that the man using it could often reload before his opponent could rush him.

This is an extremely important development. I have already described how, with ancient muskets that were slow and clumsy and inaccurate, infantry had to keep in close order; if they did not do so they would be ridden down by cavalry or charged by the opposing infantry when reloading and helpless. But as reloading came to need less time, and as it became more and more dangerous to approach within some hundreds of yards of an infantry trained to aim, there obviously had to come a period when this need was relaxed and some part of any infantry formation, or perhaps the whole of it, could operate in open order, relying on its fire-power and its power to move to protect itself against enemy action. This period had been reached before 1776, when Washington's straggling militia and half-starved 'Continentals' faced good British and German troops, and to the world's surprise dealt harshly with them.

For generations, even perhaps for centuries, it had been of advantage to a body of infantry that when the forces were in close contact the enemy should fire first. 'Gentlemen of the Guard, fire first!' was not a mere courtesy; there was solid advantage to be gained from it. In the thick smoke of the enemy's volley, and while his men were reloading, you could go forward to fire and then to charge. But now the enemy, if treated in this courteous and conventional manner, was able to open fire from a considerable distance and repeat their fire before you could get at them. Soon it was of great advantage to get your shot in first. And therefore soon there was room for skirmishers on every battlefield.

But armies have a brittle conservatism which forces innovations to take roundabout routes. Militia or volunteer forces made up hastily from civilians can more easily do the com-

monsense thing, or adopt the obvious tactics. Skirmishing and
skirmishers had become technically possible well before the
American War of Independence. But such methods and such
men were wholeheartedly condemned by all decent soldiers as
cowardly and unfair. That they could be effective was proved
by the battle of the plains of Abraham near Quebec, in 1759,
when 'a great many of our officers and soldiers were wounded
by a body of burghers from Quebec, selected as good marksmen, who lay concealed in a field of corn opposite to our right.
It was from these skulkers that General Wolfe received both
his wounds, as he gave direction in the front of the line.'
(From the *British Magazine* March 1760, quoted by Field,
*Echoes of Old Wars*.)

These 'skulkers' were soon to be paralleled by others whose
fire would not merely kill a general at the moment of his
victory but would defeat whole armies. A British officer describes the effect of the guerrilla tactics of the first American
Colonists to fight, at and near Lexington, as follows: 'The
Country by this time took ye Alarm and were immediately in
Arms, and had taken their different stations behind Walls, &c.,
on our Flanks, and thus were we harassed in our Front, Flanks
and Rear ... it not being possible for us to meet a Man otherwise than behind a Bush, Stone hedge or Tree, who immediately gave his fire and off he went.' (Field, *Echoes of Old
Wars*, a letter from Major W. Soutar of the Marines, 22 April
1775.)

It is so natural for the modern infantryman to think in
terms of cover that he needs to be reminded that, for hundreds
of years, the main engagement between armies took place outside cover, or with cover for only one or two positions held by
units of troops. After the battle of Bunker Hill, which began
the American War of Independence, the adjutant of a British
Marine Battalion wrote home, 'We killed a number of rebels,
but the cover they fought under made their loss less considerable than it would otherwise have been.' (Field, *Echoes of Old
Wars*.) It is now generally recognized by historians that this
engagement, although technically the British remained masters

of the position, was a victory for the Americans, whose fire-power and use of cover caused casualties so heavy that the British force was partially crippled thereby.

Another British officer writing of the same battle said: 'The ground ... is the strongest I can conceive for the kind of defence the rebels made which is exactly like that of the Indians, viz. small enclosures with narrow lanes bounded by stone fences, small heights which command the passes, proper trees to fire from and very rough and marsh ground for the troops to get over.' (Field, *Echoes of Old Wars*.) At a later stage in the campaign British forces found the normal type of Colonial fence to be a very serious military obstacle. These fences were made by men clearing the ground of trees, and consisted of tree trunks piled in an open zig-zag so that the ends locked. Thus at the beginning of open order infantry fighting we find the first development not only of cover, but of the use of field fortifications covering most of the front of an army in actual battle. For some centuries field fortifications had covered much of the front of armies extended in cordons through the Low Countries, the cockpit of Europe. But these armies had usually come to fight outside their entrenchments and battle had occurred only when, by piercing a line, or moving to a flank, the opponent could be caught without protection. Now we enter a period when the fortifications of the Battle of Monmouth change gradually into Lee's breastworks in the Battles in the wilderness of Virginia and finally into the trenches of 1914–18.

Washington's army contained several units, such as Morgan's Rifles, that consisted largely of men who had earned their living by hunting. They carried 'squirrel guns', weapons accurate enough to shoot a squirrel. They made ideal skirmishers; and their independence and initiative were as important as the weapons they carried. They required no drill masters; it was not necessary for the drums to beat a 'flam' before they fired. Their rifled weapons were incomparably better than the Tower muskets of the British; even the ordinary American muskets were better than this old weapon, which could seldom hit at forty yards or more. For instance, the yellow flints in the British muskets were good for only fifteen rounds, the American black flints were good for sixty rounds. Washington's army was not made up only of skirmishers and militia. Out of his own relatively long service soldiers he hammered Regular units of the line capable of the concentrated volley fire, and shock action and co-ordinated close-order movement, that would enable them to act as a solid striking force.

When a new tendency appears in human history, it often begins by showing in miniature almost all the tendencies that will matter in the development of the next period of the future. Thus here at the beginning of modern long-range fighting we have the skirmishing, the fight by fire from open order, the use of cover, the use of field fortifications covering most of the front of an army in action, accurate rifle fire, and – most important of all – an example of close combination between a Regular striking force of well-trained troops and the efforts of a militia little more than an armed population in quality. I have emphasized the value in this war, and the effect on future wars, of these 'skulkers' whose fire largely destroyed British power in America. But it is necessary also to emphasize that with this type of soldier only, Washington could do little; the British were able to occupy Philadelphia and Washington's force was condemned to the heartbreaking miseries of Valley Forge. He needed, in addition to his skirmishers and riflemen, a force capable of standing up to hammering. He made this force partly by employing a Prussian drill-master, Von Steuben, trained in the schools of Frederick the Great, and partly by gaining the direct assistance of good French troops.

All the same it was not a Prussian-drilled or French Royalist force that drove the English and Hanoverian soldiers out of the United States; it was a force that combined the drilled technique of Europe with something new. And the French officers who came back from America were the first to introduce into Europe what had been learned there about skirmishing.

In the American War of Independence, then, were almost

all the factors of the Napoleonic period of warfare. But one essential factor was missing: the field gun. Few guns were used on either side, and they had little decisive tactical important. Napoleon's remaking of warfare, on the other hand, consisted essentially of a fuller development of the new elements in war that had been displayed in America and, in addition to these, a new use of fire, particularly of artillery fire, to open the way for decisive manoeuvre.

Two facts reveal the coming of the Napoleonic change in war. One is that the decisive battle of Valmy, in which the French Revolution at its weakest and least prepared defeated the most dangerous threat to its existence, was in the ordinary sense of the word scarcely a battle at all; it was a cannonade. Infantry and cavalry scarcely made contact at effective ranges; artillery did so, and decided the issue. The second fact, of perhaps lesser importance but very revealing, is that Napoleon himself achieved power in France partly by the use, and more by the threatened use, of his own political formula: 'a whiff of grape-shot' in the Paris streets. There have been plenty of cases of captains or generals seizing power by the threat of action by the troops they commanded. Napoleon was the first, I think, to win power by something that was becoming more important than the command of legions—the command of a technical instrument, artillery.

Napoleon had not the smallest interest in tactics. That may seem a sweeping statement to make about a great captain who revolutionized warfare. But it shows in all his work, in his battles as well as his writings. It is proved by the fact that the French tactical regulations, the *Règlements d'Infanterie*, drawn up under the monarchy in 1791, were never altered by Napoleon. As first Consul or Emperor he altered almost everything else: the law of property, the division of France into Dèpartments, all sorts of civilian things. He never altered the regulations for the training and use of his troops. They survived until the reign of Louis Philippe, many years after Waterloo. Oman comments: 'This is all the more strange because that compilation (the *Règlements*) was singularly deficient in the section dealing with skirmishing and the use of light troops. It had the three-deep Frederician line and the column of companies or divisions as its base, and knew nothing of the attack by dense swarms of *tirailleurs* (skirmishers) which had been the salvation of France in 1793.' (Oman, *Studies in the Napoleonic Wars*, p. 91.)

Napoleon did not need to worry about tactics. He had ready to his hand the new tactical instrument formed by two tendencies: recognition of the value of skirmishing by all technically progressive officers, on whom the American War had impressed this main lesson, and the Revolutionary heavy column determined to break through the enemy. To these two main factors he added, in the form that we shall see, the new powers of artillery.

During the eighteenth century, with the improvement of the musket, fire became so important an element in battle that all infantry was normally drawn up in the formations most suitable for their fire: long lines of men three or four deep. Shock still entered into battle, but fire had become so important that tactical formations were designed first for fire and only secondarily for shock. The essence of the change brought about by the French Revolution and Napoleon can be considered as this: 'division of labour' began, under which fire was carried out by skirmishers in extended order and by concentrated artillery, while shock was carried out by columns of troops disposed in the most suitable order for shock – roughly a square mass, rather than a thin line. This 'division of labour' developed out of the inescapable needs of the French Revolution: with disorganized and ragged armies, raised hastily and scarcely trained, it had to face a world armed against it. These armies could not stand in line, were not drilled for the meticulous tactics of Frederick and his successors. They had courage, however, and they had numbers. They had courage from their beliefs, and numbers from the levy *en masse*. The Revolution swept aside all barriers to the production of arms; men swarmed into the workshops demanding the right to make the instruments of war; those controlling these workshops were often forced to open them to general public inspection, or even to put the forges out in the Paris streets, to satisfy the public that they were doing everything possible to make the necessary weapons. With these weapons the Revolution armed two or three times the numbers of the armies threatening Paris. Oman writes: 'The new tactics of the Revolutionary army were evolved from a consciousness of superiority in this respect (numbers), a determination to swamp troops that manoeuvred better than themselves by hurling preponderant masses upon them, regardless of the loss that must necessarily be suffered.'

And Oman defines the tactical system of these armies, which in its most developed form was taken over by Napoleon, as follows:

It consisted in throwing at the hostile front a very thick skirmishing line, which sheltered and concealed a row of columns of the heaviest sort. The idea was that the first line of *tirailleurs* would so engage the enemy and keep him occupied, that the supporting columns would get up to striking distance with practically no loss, and could be hurled, while still intact, upon the hostile first line, which they would pierce by their mere impetus and weight, since they were only exposed to fire for a very few minutes, and could endure the loss suffered in that short time without losing their *élan* or their pace. The essential part of this system was the enormously thick and powerful skirmishing line. Whole battalions were dispersed in chains of *tirailleurs*, who frankly abandoned any attempt at ordered movement, took refuge behind cover of all sorts, but were so numerous that they could always drive in the very thin skirmishing line of the enemy, and get closely engaged with his whole front. The orderly battalion fire of the Austrian or other allied troops opposed to them did comparatively little harm to these swarms, who were taking cover as much as possible and presented no closed body or solid mark for the volleys poured at them. There is a very clear description of such a fight in Ditfurth's narrative of the Battle of Hondeschoote, where Walmoden's Hanoverians, covering the Duke of York's flank, fought for four hours against a swarm of *tirailleurs*, who always gave way and took refuge in hedges or buildings when attacked by the bayonet, but always came back to molest the defensive line opposed to them, till after clearing their front eleven times the Hanoverians had to give way in the end, because their original three-deep line had simply been shot to pieces, and about a third of their men had fallen. (Oman, *Studies in the Napoleonic Wars*.)

This tactical system of the French Revolutionary armies, based on their needs and moral qualities, led to another change that had strategic results. These armies moved as men hurrying: they had no very strict order to keep and they did not march in step. (Normal French infantry, to this day, straggles out of step most of the time.) But the speed of their march, into action and during action, was the quick-step of a hundred and twenty paces to the minute. The armies they fought against maintained the normal barrack-square drill step of seventy paces to the minute. The quick-step is quite natural to a civilian hurrying; the slow march, as Home Guards have found in Britain, requires long practice. The old armies marched slowly because only when marching slowly can a perfect line be maintained, in spite of any irregularity in the ground. The new French army marched swiftly because its columns, unable to fire much themselves from their narrow frontage, needed and wanted to get swiftly through the field of fire and pierce and overthrow the enemy lines. This, with another factor, gave Napoleon's armies their strategic and tactical mobility.

The chaotic supply system [writes Liddell Hart] and the undisciplined nature of the Revolutionary armies compelled a reversion to the old practice of 'living on the country'. And the distribution of the army in divisions meant that this practice detracted less from the army's effectiveness than in the old days. Where, formerly, the fractions had to be collected before they could carry out an operation, now they could be serving a military purpose while feeding themselves.

Moreover, the effect of 'moving light' was to accelerate their mobility, and enable them to move freely in mountainous or forest country. Similarly, the very fact that they were unable to depend on magazines and supply-trains for food and equipment lent impetus to hungry and ill-clad troops in descending upon the rear of an enemy who had, and depended on, such direct forms of supply. (*The Strategy of Indirect Approach*, p. 122.)

The distribution of the army into divisions, mentioned in the preceding quotation, was another development of the period before his own that Napoleon brought to its full development. The armies of the past had often been sub-divided, but sub-divided into the troops – or at earlier periods, the personal followers – of each subordinate commander. Each of these sub-divisions was likely to consist of infantry or cavalry, who might or might not have some guns with them. Only in the case of separate armies joining, as when Eugene and Marlborough joined, would each of the sub-divisions of the army be likely to have infantry, cavalry, and artillery 'belonging to it' in a useful proportion. But the formation of army corps, which began before the French Revolution, altered all this. These corps were real little armies 'on their own'. Each had its own infantry, its own cavalry for protection and reconnaissance and pursuit, its own guns and services. Each could act apart from the other corps that made up the army.

This was in its inception a development of the straggling 'cordon' system of war along the French frontiers and in the low countries, when the eighteenth-century generals spread their armies out in wide detachments. Naturally each detachment needed, if possible, to be self-contained. But it had no considerable influence on warfare until Carnot, the military leader of the French Revolution, began to organize these separate divisions in such a way that while operating separately each could, if need be, support others, or all could, if need be, concentrate at one point for decisive action. Napoleon developed this idea and expressed it in his famous slogan; 'March separately, strike together.'

The division of armies into self-contained units of this sort, and the weaving of these units into co-operative dependence on each other, had at first strategic importance rather than tactical. It mattered in bringing troops to battle rather than in their behaviour on the field of battle. But in Napoleon's hands, while maintaining the utmost strategical importance, it also acquired a tactical value. His 'divisions', separated in time and space, would be so concentrated for battle that they came not only against the enemy's front but also by the very direction of their march against the enemy's flank or rear. And Napoleon developed also a new variant of the old idea of 'fixing and hitting'; he had often, if not always, a corps capable of independent action which acted in time or space as his advance guard; it was the first to engage the enemy, pin down the enemy's main forces, and enabled Napoleon to bring his other units to decisive attack against an enemy whose forces were known in strength and position. This 'grand advance guard' did not always go in front of Napoleon's other forces; it might be to a flank; it was always the corps against which the enemy was likely to take action first.

This development of the army into four or five separate corps moving each a day's march or so away from the next, and so arranged in square or lozenge formation that concentration was possible upon any of them that might encounter heavy enemy forces, belongs more to the sphere of strategy than that of tactics. But it combined with another factor in Napoleon's technique to influence the tactics employed. He needed quick and complete victories. It was not enough for him to outmanoeuvre an opponent and force that opponent to fall back, or to meet the opponent in a simple head-on clash from which the defeated army could withdraw in good order. Napoleon had often to fight campaigns in which the number of men at his disposal was considerably less than that of his opponents. He would concentrate his corps on a section of his opponents' forces in such a way as to have, normally, superiority on the field of battle. But if he had then indulged in what might be called 1916 tactics, a long laborious fire-fight, a gradual attrition of the opponents' force, and then an assault over a wide front, his superiority on the field of battle would have disappeared before decision came. The opponent would have reinforced the section of his army engaged, from detachments scattered elsewhere. This need for a speedy decision is one of the reasons for the Napoleonic retention of fairly heavy columns in attack. The strategical method imposed the tactical method on him.

I have found that young soldiers are sometimes misled by the use of the word 'column' to describe the Napoleonic formation in attack. These young soldiers think of a column as they themselves know it: a long winding snake of men moving forward three or four abreast. Or if more modern, and mechanized, they think of a column of vehicles strung out one behind the other, a column of tanks behaving as if they were ships at sea. Actually, the Napoleonic column was exactly parallel to the phalanx. A normal battalion in the column formation used for attack would have a frontage of thirty men and a depth of thirty. Only the first two ranks of these men could use their muskets – sixty out of nine hundred. This battering ram of men depended very little on its own fire-power; it depended for most of its effect on shock. And the preparation by fire for its shock action was carried out by the skirmishing company of each battalion, one-tenth of the strength, and by artillery.

Naturally, when marching, Napoleon's troops moved three or four or six abreast, according to the width of road available, And even when fighting they did not always move in the square columns described. They were given quite often a formation called *ordre mixte*, consisting of alternate battalions in column and in extended line, three deep. This formation permitted far more fire-power, and was used fairly often by troops playing a defensive part in the battle.

The normal formation of the British infantry at the beginning of the Napoleonic Wars was the same as that of the infantry of Frederick the Great: a line of men three deep. The third rank could not fire their muskets, with safety to the ranks in front of them. And there were no skirmishers, or very few, ahead of this line during action. When fighting in this formation British troops had few successes; and it was not until Wellington developed his characteristic linear tactics that the British infantry began to show a superiority over their opponents.

After the war in America the authorities in London had, of course, not only disregarded its lessons but disbanded the formations of light troops, Rangers and Rifles, with which the British generals in America had tried to cope with Washington's marksmen. Luckily the British army possessed a really progressive soldier to train the new bodies of light infantry necessary. This soldier, Sir John Moore, not only broke with tradition in regard to uniform and drill; he also insisted on men being taught by patient explanation of the reasons for what they were supposed to do. He saw that skirmishers and riflemen could not be expected to fight well under arbitrary discipline, doing things by drilled habit rather than by understanding. He formed light units so powerful and so well instructed that Wellington was always able to put in front of his main line a heavier array of skirmishers than that covering the French columns attacking him. Oman writes of Wellington that his experience 'in Flanders in 1794 had taught him that the line cannot contend at advantage with a preponderant mass of light troops, which yield when charged, but return the moment that the charge is stopped. The device which he had thought out to provide against this danger was that he would always make his own skirmishing screen stronger than that of the enemy.' (Oman, *Studies in the Napoleonic Wars*.)

Here then is a very simple pattern of part of the development of war. The armies of the French Revolution and of Napoleon win their victories partly because they have a strong screen of skirmishes and are fighting against troops which have a weak screen or none at all. Then later Wellington's army develops a screen of skirmishers heavier than that of the Napoleonic armies, and fairly generally succeeds in defeating them. The matter was not of course as simple as all this. But we can say that infantry in open order (skirmishers) were continually becoming of increasing importance as compared with infantry in close order.

Wellington did not hesitate to break up battalions or larger units, and scatter their companies through his army as light troops. And he trained a large proportion of the Portuguese put under his command as light battalions and companies for the skirmishing lines of his divisions. An Anglo-Portuguese division of eleven battalions would have no less than eighteen companies of skirmishers ahead of it; a French division of the same size would have only eleven. In other words the British army met the new tactics of the French partly by retaining what was good in previous tactics – the firing-power of the line as compared with the inability of the column to use more than a few of its weapons – and partly by 'going one better' than the French in the development of troops fighting in open order.

But the French armies had another, and according to the argument I am pursuing, an even more important development of tactics which helped to give them victory from one end of Europe to another. Napoleon phrased this development with a clarity that does not mark all his statements on war. 'Même en plaine les colonnes n'enfoncent les lignes qu'autant qu'elles sont appuyées par le feu d'une artillerie très supérieure, qui prépare l'attaque.' (Even in flat and open country columns cannot break through lines except to the extent that they are supported by the fire of a very superior artillery, which prepares the attack). These were Napoleon's words to Foy, reported in the latter's memoirs. They embody a truth about warfare that remained a truth until at least 1917.

The student of warfare will not find it necessary for me to emphasize Napoleon's increasing preoccupation with concentrated artillery power. For the civilian, who may not have realized how much Napoleon was a gunner, it may be worth suggesting that Tolstoy's *War and Peace* should be read again. His description of Borodino, the victory that Napoleon won at the point where Hitler in 1941 failed to win, is of most convincing accuracy. And it shows the Emperor placing and giving orders to his artillery before he deploys or launches any of his troops. This was not a mere question of time, of doing one thing which comes first in time before the next which necessarily comes later. It was Napoleon's constant preoccupation, as he learnt more of war from experience, to see that his attacks were covered by so heavy a bombardment that the enemy were shaken or even half broken at the one vital point on the battlefield where decision could be obtained. And his strategic insight was such that he chose this point with unerring logic, either at the beginning of the battle or as soon as he had found out by fighting where was the weak link in the enemy organization. He massed his guns against this weak link, pushing them forward boldly, sometimes well ahead of his infantry; and in these guns he used a high proportion of grape shot or case shot. This consisted of many bullets or shot which scattered out from the mouth of the gun to form a deadly pattern.

Wellington had an answer to the French tactics of concentrated artillery fire. He could not answer it by a similar concentration, for he was seldom given the guns. No one in England had as much insight on this matter as Sir John Moore had possessed on the parallel need for skirmishers. It is significant, for example, that when Wellington for the first and last time met Napoleon in battle, at Waterloo, the British general had a few hundred more infantry than the French Emperor; but he had only 156 guns against Napoleon's 246 guns. So Wellington developed another way of meeting the French attack: instead of 'going one better' along the same lines, he hid his troops from artillery fire. By doing so he opened a new era in the tactical handling of infantry.

All through the past of warfare it had been natural for generals to choose, when they could, a position for their armies from which the ground sloped down towards the enemy. They had in this way 'the upper hand'; in any charge or shock action their troops would have the advantage of moving downhill. The enemy climbing towards them were physically exhausted by the effort of climbing before they came to action. In these old battles, when no long-range weapon could damage troops exposed to view, there was no reason to hide an army or any considerable part of it; indeed most commanders liked to display all their forces in order to overawe the enemy. But now a long-range weapon was becoming of great importance. Wellington avoided its power by choosing positions behind or beyond a rise of ground, in such a way that his troops, often kept lying down until the moment of action, were hidden from view and therefore from fire. This gave him not only tactical security against one of the most powerful factors in the French attack; it also gave him the chance for tactical surprise. The French generals could seldom find out exactly where his line extended. In several battles they pushed their columns forward on a flank in the belief that they were outflanking him; his hidden line was beyond them and took their columns from both flanks.

It was only when absolute necessity compelled, owing to no cover being available in some parts of the line, that Wellington occasionally left troops in his battle-front visible to the enemy, and exposed to distant artillery fire. The best-known instance of this was the case of his centre brigades at Talavera, which were unmasked perforce, because between the stony hill, which protected his left, and the olive groves, which covered his right, there were many hundred yards of open ground without even a serviceable dip or undulation in which the line could be concealed. And this was almost the only battle in which we find record of his troops having suffered heavily by artillery fire before the clash of conflict began. (Oman, *Studies in the Napoleonic Wars*, p. 101.)

These words, by the great historian to whom I owe so much of my material on the Middle Ages and on the Napoleonic period, explain his judgement that, against English troops handled in Wellington's way, 'it was the French tactical formation which made (a French) victory impossible.'

Strategically, and on a world scale, Napoleon was a far greater soldier than Wellington. But Wellington was far more than 'the full stop at the end of the chapter', as he has been called by an English historian. He provided, on the scale of tactics, the two answers that fact and action proved effective to the system of warfare introduced by the American and French Revolutions, the system brought to world-shaking development by the strategic and political genius of Napoleon. And he played his part in carrying war to an even higher stage of development, a stage that Napoleon could not understand. This stage consists essentially of a combination between a well-trained Regular striking force and an active armed population. Napoleon came up against this new level of warfare to some extent in Russia, and to some extent in Germany, where serfs were freed and far-reaching reforms carried out in order to form an army, and a *Landsturm* or Home Guard, that would be national in character rather than servile. But it was in Spain that Napoleon met the full force of this new level of warfare. And it is correct that the word describing the methods of fighting of the armed population should be a Spanish word, *guerrilla*, 'little war'.

Guerrilla warfare is all skirmishing. It is all 'Indian fighting'. It combines the most primitive methods and tactics with the new factor that during the Napoleonic period is helping to revolutionize war: the individual weapon accurate enough to hit an enemy individual before that enemy sees the man who is handling the weapon. The 'Spanish ulcer' that drained the strength of Napoleon's empire consisted, technically and tactically, of the new way of fighting developed first in America, then by the French Revolutionary armies, then by Wellington, detached from the old way of fighting: guerrillas are skirmishers operating 'on their own' away from – yet in necessary combination with – the regular striking force. The individual fighting man, no longer subject to drilled movement in action and no longer controllable immediately and at all times by a superior, takes his place in decisive action not only as an auxiliary fighting a few hundred yards ahead of the main body of the army but also as an auxiliary operating many miles away, against the weakest part of the enemy's organization, his lines of communication. This is an essential factor in modern warfare that we shall see disappear and reappear.

Tactics, from the development of the musket and bayonet to the armies of Frederick the Great, had mainly followed the line of making a strong disciplined body of drilled infantry whose fire-power and shock-power depended on their automatic obedience. Napoleon recognized the change that came with his own era when he said to Jomini, 'I do not love armies of automatons, I require soldiers to be intelligent.' It was still possible for a single man, if he was Napoleon, to take all the strategical and most of the tactical decisions necessary governing the movements into action and during action of armies of decisive size. Napoleon did this; his Chief-of-Staff was a mere secretary formulating and distributing his orders. But when the process of splitting up armies passed from a formal organization into army corps or divisions into the more radical 'division of labour' between striking force and guerillas, Napoleon and his marshals no longer had a method of command and control suitable for this new shape of warfare. The history of the Peninsular War is in essence a history of the French army dispersing to destroy guerrillas and then being struck when dispersed, or threatened, by the regular forces of Moore and Wellington. As soon as these French armies concentrated to check or throw back the British striking forces, the guerrillas rose again and forced them to a new dispersal. This was a form of war to which the Napoleonic system could find no effective answer.

In Frederick's day the main process of battle had been preparation by musket fire, with some assistance from artillery, and then assault in line with the bayonet. In Napoleon's era the main process of battle was preparation by concentrated artillery fire, with some assistance by skirmishers and other musket fire, and assault in column with the bayonet. But towards the end of Napoleon's day there came these new developments that he could not meet; Wellington's concealment of his main body behind strengthened skirmishers and natural cover, and the guerrillas' concealment in every patch of difficult country or territory lightly held across his flanks and rear. The day of invisible armies had begun, the period symbolized for us by the medieval complaint that knights should fall 'almost without seeing the men who slew them'. Industry was producing weapons of such range and accuracy that armies were beginning to hide themselves in defence and to attack swiftly, by surprise or stealth, or behind a cover of gunfire and skirmishes, so that the attacking troops should be exposed to fire for the briefest possible time.

All weapons and tactics have been changed since Napoleon's day; but many of the tendencies in war that first developed in those days continue into the present.

# 8. Machine-Gun and Trench

A striking phrase occurs in one of the histories of the American Civil War. Describing a manoeuvre by Grant, and Lee's counter to it, the historian writes: 'Lee manoeuvred his fortifications....'

The history of tactics during the period after Napoleon is partly the history of the development of fire-power, which at first loosens and speeds up warfare and later ties it down to relative immobility; it is partly the history of the development of field fortifications, which reach their full development in the Hindenburg Line of 1917 and are brought to useless perfection in the Maginot Line of 1939. But whereas the development of weapons can be shown as a rising curve on a graph, the development of fortifications for a field army came to their limit in one form, then almost disappeared, to reappear later in another form.

In the previous chapter we mentioned the American use of the heavy timbered Colonial fences as breastworks. During the American Civil War the decisive area between the two capitals was so heavily wooded that breastworks of tree trunks could be constructed along almost any position chosen for defence. Lee, whose army during the decisive campaign was on the defensive, made continual and increasing use of such breastworks; and Grant's artillery was never of sufficient power to breach these field fortifications effectively. They therefore had to be assaulted, and battle became largely the question of the power to storm or to assault some 'Bloody Angle' of logs behind which the defenders were sheltered from musket and from rifle fire. Assault had become more difficult by the time of the American Civil War than it had been in the Napoleonic period. It was more difficult for the attacking army to push its field guns forward, in front of its infantry, before the assault began, and to use these guns at relatively short range to smash a hole in the enemy's formation. Sniping and skirmishing by men armed either with rifles or with accurate muskets killed so many gunners that they could not remain exposed to enemy view in front of their own main line of infantry; they had to take position behind it. The distance to be covered by assaulting infantry under fire had increased with the increased range of weapons. And finally, and most important, the greater speed with which small arms could be loaded and fired made it certain that defending troops could still hold their positions with fire, not with bayonets, even when the assaulting troops had got close up to them. The bayonet, the last shock weapon left in warfare, was ceasing to have actual fighting value. General John B. Gordon of the Confederate Army wrote in his *Reminiscences of the Civil War*: 'Very few bayonets of any kind were actually used in battle.... The day of the bayonet is past except for use in hollow square, or in resisting cavalry charges, or as an implement for constructing light and temporary fortifications.' The view that the bayonet was now of little value was confirmed by Surgeon-Major Hart and J. E. B. Stuart's Chief of Staff (see General Fuller's *Dragon's Teeth*, p. 245).

It was becoming increasingly impossible for formed bodies of troops in close order to charge across the field of fire separating armies and reach the enemy's position while still in sufficiently close formation to use a shock weapon. This was not a question of morale or discipline; at battles such at Gettysburg or those of the Wilderness campaign, troops continued to advance after receiving very heavy casualties. But it was physically impossible for them to advance in close ordered line or column; the enemy's fire destroyed these formations and reduced them to straggling groups of skirmishers capable of fire action but not of shock action. Lee's field fortifications, wooden breastworks of sufficient height to be an obstacle to men charging, fulfilled two functions: they gave shelter to the defending forces and they physically checked the attackers at points where the defenders' fire would wipe them out while they were checked. Ordinary trenches do not fulfil both these functions; they give shelter from fire but do not provide a physical obstacle to the assault. It was not until modern industry had so developed that thousands of miles of barbed wire could be coiled in front of the trenches that field fortifications became again of such importance that assault against them was a matter of the utmost difficulty. Between these two periods, that of the breastwork and that of the wired trench, lies a gap in which field fortifications were of lesser importance, though they bulked largely in the Crimean War. And few soldiers were willing to recognize the decrease in value of the bayonet, or the fact that men in close order could no longer move under fire.

The difficulty of assault, of a charge by unarmoured foot-soldiers against the fire of precision weapons, was hidden from most students of warfare by a myth and by a reality misunderstood. The myth was that of 'cold steel'; faced by the increasing complexity of modern weapons and tactics, conservative commanders retreated morally and mentally to the shock weapon that they could understand, that seemed both 'safe' and heroic, and was intimately connected with the social myth by which the military castes in Europe maintained their political power and economic security. This social myth, gradually developed over a considerable period, included the proposition that courage, discipline, will-power of peculiar and exclusive sorts had to be possessed by the officers leading troops in modern battle; the need to cultivate these qualities during a whole lifetime was the justification for maintaining an exclusive caste of professional officers. This myth came to pieces during most real wars; the American Civil War was largely fought by 'civilian' leaders, and the French fought best in the War of 1870 when their emperor and generals had surrendered. But the myth was sedulously rebuilt during the periods between wars; in Britain it even survived to a quite large extent the successes of unprofessional Boer farmers, and even outlasted the war of 1914–18, in which officers who had not been professional soldiers had perforce to be used (after the first year) for most of the actual fighting, while officers who had been professional soldiers were to a considerable extent – usually against their will – translated to the unheroic position of command from a considerable distance in rear. This myth preserved the bayonet, as it preserved the purely decorative sword and lance, because these officer castes felt that the bayonet charge was their special accomplishment that no civilian force was likely to attempt.

A misunderstood fact also kept the idea of the bayonet alive. The fact was that the Prussian army in six weeks of ceaseless advance had scored a decisive victory in 1870. Soldiers admiring this feat of arms misunderstood the essence of the Prussian tactics. Although Colonel Henderson wrote in his *Science of War* (1905) that 'the Germans relied on fire, and on fire alone, to beat down the enemy's resistance; the final charge was a secondary consideration altogether' – in spite of this plain statement by the best military theoretician in Britain at that time, the British and other armies continued to believe that the Germans had won their victory by a fire-fight leading up to assault. They ignored entirely an important fact: that all German infantry formations moving forward in action during that war broke up into loose groups of skirmishers, greatly to the dismay of many German officers who insisted that they ought to move forward in close columns of half companies. And they also ignored the fact that the Germans had won this war by the strategic offensive combined with the tactical defensive: German columns pushed forward to threaten flank or rear of French forces in such a way that the French were usually compelled to try to attack and clear the Germans out of these positions. This was the typical shape of the battles up to and including Sedan.

It is difficult to get any serious measure of the value of the bayonet during the Franco-Prussian War, but in the present century casualty statistics have been improved to such an extent that it becomes possible to judge statistically the relative value of various weapons. In the Russo-Japanese War of 1904 about

two and a half per cent of the total casualties on both sides
were caused by spears, swords and bayonets. In the Great War
of 1914–18 bayonet wounds became so rare that no full statis-
tical record of them was kept. They are included, in British
figures, among the 1·02 per cent of miscellaneous casualties
and accidents. But just before that war the author of this book,
like any other Englishman then being taught any form of infantry training, was taught that attack was carried out by
advance in extended order, by short rushes when the enemy
fire was considerable, by the building up of a firing line and the
mastery of the enemy's fire, and then by a bayonet charge,
which should be accompanied by loud yells. This, as the basic
pattern of infantry attack, was already at that time fifty years
out of date. But it continued, and in some benighted spots still
continues, to be taught. One result of it is that after every
campaign – Norway, France, Crete, Singapore, Libya – newspapers report that British troops 'complain they can never get
at the enemy with cold steel'. Of course they cannot; they get
killed first.

The development of the machine-gun in the period between
the Franco-Prussian War and the War of 1914–18 increased
the difficulties of infantry attack and made the defensive,
tactically at least, very much the 'strongest form of war'. Fire,
as it developed through the centuries, had for a time made
tactical movement easier, because its concentration against
enemy units exposed in the open, and closely grouped to give
good targets, enabled holes to be blown in the enemy's position.
But now the effect of fire was reversed. Enemy armies hidden
behind cover could weave in front of themselves so closely
linked a web of fire that the attacking infantry found it hard to
get through. And early in 1915 it began to be generally
recognized that machine-gun fire could 'lock the front'.

The development of the machine-gun before 1915 was relatively slow. That development begins for practical purposes in
1882 when an American inventor, who had concentrated on
electrical and chemical processes, met in Vienna an American
business man whom he had known in the United States. The

business man said to the inventor: 'Hang your chemistry and electricity! If you want to make a pile of money, invent something that will enable these Europeans to cut each other's throats with greater facility.' The inventor was Hiram Maxim, and this chance conversation started him working on the development of the first modern machine-gun.

Long before Maxim's time, men had tried to make a machine that would fire projectiles rapidly one after another. When the Spaniards were developing the first good firearms, a Spanish commander mounted a whole row of these firearms on thirty carts, which he put in front of his infantry. This commander, Pedro Navarro, had several men in the carts whose duty it was to fire all the arquebuses at once, by means of a quick-match or fuse that connected all the weapons in each cart.

In 1693 a ten-shot repeating musket was reported; but something must have gone wrong, for it was never heard of again. Later, six to ten muskets would be linked together on a single frame, and this frame would be dragged into action on wheels. These clumsy weapons were called 'organs', because they looked like these musical instruments. In 1718 a gentleman named Puckle patented an elaborate machine-gun with revolving chambers; special chambers could be used for firing square bullets at Turks, since round bullets were too humane for the killing of infidels.

A more effective gun was used in the American Civil War, that invented by Dr Gatling of Chicago. He had six barrels fixed round a central axis, as if they were staves of a cask. The six barrels revolved about the axis; as each barrel came to the top a cartridge fell into it from a trough. By the time this barrel had come to the bottom of its movement, the cartridge had been pushed home and the breech closed. The cartridge was fired and the barrel began to come up again, ejecting the empty cartridge case. This type of gun was later used by the British army in colonial wars. The barrels were rotated by a man turning a handle at the side of the gun.

Some years earlier a Belgian officer had invented a machine-gun with twenty-five barrels, each of which could be fired twelve times a minute, throwing out three hundred bullets in a minute. This gun weighed over a ton, and was carried on artillery wheels as if it had been a normal piece of artillery. In the Franco-Prussian War of 1870 the French foolishly tried to use it as if it were a field gun. As it had less than half the range of the Prussian field guns, and as the idea of hiding guns had scarcely been thought of, this French *mitrailleuse* was of very little use; it was usually knocked out by Prussian field artillery shells before it could do much damage to the Prussian infantry.

After the failure of the *mitrailleuse* few soldiers believed in the machine-gun. In 1873 a Swedish banker, Nordenfeldt, took out a patent for a gun rather like the medieval 'organ'. in 1874 an American, Hotchkiss, patented a weapon rather like the Gatling. Little real progress was made by these inventors until a man of quite a different type came on the scene – Basil Zaharoff.

Zaharoff was a salesman for arms, friend of financiers, adviser to war ministers, and before his end the 'mystery man of Europe' and a power as great as ministers. Zaharoff was not a mere profiteer. He sold the best weapons. When he had sold them to one country, he proceeded to sell rather better weapons to a rival country.

It was not of course his business to promote wars. But his business flourished when wars happened to promote themselves. And he, rather than Maxim or the other inventors, should receive our thanks for the machine-gun.

He had been selling very light unarmoured torpedo-boats to various governments. It was suggested to him that machine-guns would be good weapons with which to protect larger vessels from these little torpedo-boats. Zaharoff adopted the idea with enthusiasm. It is much more profitable to sell a weapon, and an antidote to that weapon, than it is to sell the weapon alone.

After tests at Portsmouth the British Government bought, for the defence of its warships, several ten-barrelled Gatlings, a five-barrelled Nordenfeldt, and several other types of machine-gun. And Basil Zaharoff received commission on these sales; he also at once received orders from various governments for rather more powerful armoured torpedo-boats, which would not be damaged by the machine-guns that the British Admiralty had bought.

Then came Maxim. As a child of fourteen, when he had first fired an army rifle, he had held it clumsily and the 'kick' of the rifle had bruised his shoulder. He remembered this. He started to make a machine-gun that would employ this 'kick' to do the business of opening the breech, ejecting the cartridge and pushing a new cartridge into the breech. Then springs would close the breech and the gun would fire again.

In all its principal features the machine-gun he invented is that of 1914 and today. The first gun he made fired ten shots in a second. Soldiers from all over Europe became interested in the gun; the Tsar's army bought it; Kaiser Wilhelm exclaimed: "This is the only machine-gun.' And Basil Zaharoff adopted it – which was more important than the approval of any war-lord.

Maxim's experiments had been carried out in London. But the British Government was the last to treat machine-guns for land warfare as important. The British army went to war in 1914 with fewer machine-guns than the Germans had; and Mr Lloyd George had to override military opposition in order to get large numbers of the guns manufactured.

According to Brigadier-General Baker-Carr, the first commandant of the British army's machine-gun school in France, many British battalion commanders of the 1914 vintage 'frankly and cordially disliked' machine-guns. He wrote: 'What shall I do with the machine-guns today, sir?' would be the question frequently asked by the officer in charge of a field day. 'Take the damn things to a flank and hide them!' was the usual reply.

Naturally, on manoeuvres, when machine-guns were treated in this way it was easy to imagine that infantry could get through machine-gun fire and 'close with the enemy' – to use the phrase that still describes, according to the War Office, the duty of British infantry. When battle came this proved to be no more than a phrase. And when the barbed wire was added to machine-guns, and the trench systems grew into line upon line of muddy ditches, tactics bogged down in futile immobility.

For many years before this the power of fire had been so clear that many soldiers had realized that frontal attack would be difficult, and had concentrated on teaching the value of flank attack. This teaching culminated in the Schlieffen plan, the scheme by which the mass of the German army would march round its enemies, through Belgium and along the coast and beyond Paris, always attacking from the flank and always pressing the French forces together, until a new Cannae or a new Leuthen had been achieved. But Schlieffen's plan was watered down by the timid old general kept in charge of the German army (largely by the social myth of which I have already written). Von Moltke the Elder had conquered Austria and France in 1866 and 1870; therefore Von Moltke 'the Younger' had to be allowed to keep nominal control of the German army in 1914; real control slipped from his nerveless fingers, and the Marne ended the Schlieffen plan. Then for a few weeks, wearying of frontal attacks that cost lives but gave little results, each army sought to outflank the other, until to the surprise of each they found that no flanks were left, that the trenched front stretched unbroken from Switzerland to the sea. The defensive power of fire, of wire, and of the spade, had ended mobility in war.

It is curious the way in which a new development in warfare can come into being on a small scale and by accident, at some critical point, and then drop back out of existence for a period. The tactical factor that, many years later, was going to end the stalemate caused by the machine-gun, wire and trenches, showed itself just before this stalemate supervened. That tactical factor is manoeuvre by vehicles. The crisis of the Battle of the Marne was marked by the first large-scale manoeuvre by petrol-driven vehicles known to military history. General Gallieni, commanding the French forces in and around Paris, moved part of these forces on to the flank of General Von

Kluck's advancing army by commandeering the taxicabs of Paris, and other vehicles. More than anything else this manoeuvre decided the Marne, the 'miracle' that saved Paris and probably saved France from defeat in 1914. Yet it was not repeated, and could not be repeated when the armies ceased to have flanks; no vehicles were available that could wrest out a flank by main force, by smashing through some part of the enemy's line. And there were not enough petrol-driven vehicles, in those days before mass producton was developed, to make possible long-range manoeuvre based mainly on the petrol engine.

The soldiers on each side of the trench lines in 1915 had therefore to seek for a solution to the problem of fronts locked by fire and fortification. They first sought it mainly in the use of concentrated artillery, and particularly of heavy artillery. Changes in the construction of guns had made it possible to bring into action, in slow-moving war, guns of a size greater than any except those old bombards of the far past used in the siege of cities. The big howitzer, firing a ton or so of metal, had been added by the Germans to the equipment of their field armies.

The essential way in which these guns differed from those of past centuries was that the recoil was mechanically absorbed by a system of springs or compression chambers or friction brakes which were part of the gun itself. From the first mortars, on their immense wooden platforms, to the field guns of Napoleon's day or of the Crimea, the recoil due to the firing of the gun caused the whole weapon to move backwards. Before it was fired every member of the crew had to be clear of the whole area into which the gun might jump. And as soon as the gun had been fired it had to be moved back to its original position and resighted. This told against speed and accuracy.

The guns of the past were fired by black powder, which made such a smoke that on a still day the gunners had to wait for a moment or two after firing their weapon and running it back into position before they could aim it again. The develop-

ment of smokeless powders and of more powerful propellants
had cut out this difficulty, and had also made it possible for
gun positions to be relatively invisible by the enemy.

But the development that really speeded up artillery fire was
that of the breech-block. A gun loaded by the muzzle was
difficult and slow to sponge out and to load and fire; and the
gunners had to expose themselves in front of their weapon for
each round. As breech-loading came in it became possible for
artillery to be worked by gunners who remained in relative
safety behind shields attached to their weapons or behind
earthworks. And the process was far swifter, so that a field gun
could pour out several shells a minute as compared with one
shell every few minutes.

With this improved artillery, many soldiers believed in 1915,
it should be possible to do what Napoleon had done in the
past: blast a hole through the opponent's line and then send
forward infantry (or even cavalry) through the gap and round
the flanks on each side of it. Most of the history of the War of
1914–18 is the history of the failure of this idea. Guns and
howitzers could be made that would, when concentrated,
slowly churn up the earth and destroy the trenches in which
defending infantry were hiding. One answer to this was the
deep dug-out such as those made by the Germans under the
Hindenburg Line. But in the main the heavy artillery of that
war provided its own answer. The aim was penetration of the
enemy's line; the guns could smash up earth and trenches, but
in doing so they made an impassable barrier in front of themselves, a barrier of torn fields in which the drainage was destroyed, of shell-holes and ruined paths, through which it was
difficult for the infantry to penetrate and quite impossible for
the artillery itself to get forward. These guns were siege
weapons, and this warfare was often treated as siege warfare.
But in fact it differed from a normal siege because new 'walls'
could be created more easily and quickly then old 'walls' could
be knocked down. Behind each 'breach' made in a defensive
system, in this trench warfare, new lines of trenches could be
dug and manned, before the attacking force could get its guns

up over the desert of mud and ruin that they had themselves created.

. The guns had been improved, but the means of transporting them had not. No pneumatic tyres existed large enough or strong enough even for the field guns. The larger heavier weapons, with steel tyres on their wheels, would knock any road to pieces even if moved only at the pace of the slowest carthorse. Howitzers used to be dragged forward by steam traction engines. And on the Western Front men more numerous than any army Wellington ever commanded spent their lives, through years of warfare, remaking roads that the guns destroyed with their wheels almost as much as with their shells.

Surprise dropped out of warfare. Every attack was a frontal attack, and the probable place and date for it was given away by the artillery preparation. And no attack could be concentrated on a narrow front to achieve a break-through, partly because the immense concentration of munitions required had to be brought up near the line and stored in dumps; as these dumps spread out from each railhead laterally, they could only feed the attack on a relatively wide front. The railways were the main means of transport for supplies; motor transport was used for only a few troop movements, and for the distribution of supplies between railhead and the dumps. Horse transport was still standard both for infantry and guns. Armies were closely tied not only to their railways but to the great dumps of ammunition which these fed. It was not unusual for these dumps to include about ten thousand tons of ammunition. If the army desired to move forward or backward it would take days to lift such supplies and transport them. Very occasionally they were lifted much more rapidly, by enemy air bombardment or by accident. Nine thousand tons of shells went up at Audruicq in 1916, and ten thousand at Saigneville and ten thousand at Blangies during 1918. Before an offensive, in this form of war, it might be necessary to create a dozen dumps of this size. The amount of ammunition expended in 1916, in one month's bombardment on the Somme, was 148,000 tons. By

1918 even more incredible amounts of steel and high explosive were being poured out. The artillery preparation at the third battle of Ypres lasted for ten days and was carried out by over three thousand guns, of which a thousand were heavy. This was an average of one gun to every six yards of front. Four and a quarter million shells were fired, costing twenty-two million pounds. Four and three quarter tons of shells were thrown on every linear yard of front.

And the result? The result was that at great cost of lives some square miles of swamp were gained; this swamp had been made impassable for guns or tanks by our own shells, and almost impassable for troops.

The war of 1914–18 has been discussed endlessly; the offensive we have just described has been one of the battles most bitterly fought in books and newspapers, perhaps as bitterly fought as on the ground. Mr Lloyd George bases much of his criticism of Haig on this costly and futile offensive. Now that we are in a very different sort of war it is easier to see 1914–18 in perspective; it is easier to see that there was an element of the inevitable about these suicidal massacres in the mud. It was not any general's fault that they occurred; they were not due to the prejudices or mistakes of a Staff or a caste of officers. They were due to the whole shape and nature of war as war then existed, a shape and nature imposed on war by the natural development of weapons. And behind that development of weapons were all the swift changes of the Machine Age, the changes in industrial technique and social life. The effect of these changes is described by Winston Churchill, in words that I have quoted before but cannot better:

Wealth, science, civilization, patriotism, steam transport and world credit enabled the whole strength of every belligerent to be continually applied to the war. The entire populations fought and laboured, women and men alike, to the utmost of their physical destructiveness. National industry was in every country converted to the production of war material. Tens of millions of soldiers, scores of thousands of cannon hurled death across the battle lines, themselves measured in thousands of miles. Havoc on such a scale had never even been dreamed of in the past, and had never proceeded at such a speed in all human history. To carry this process to the final limit was the dearest effort of every nation, and of nearly all that was best and noblest in every nation.

But at the same time that Europe had been fastened into this frightful bondage, the art of war had fallen into an almost similar helplessness. No means of procuring a swift decision presented itself to the strategy of the commanders, or existed on the battlefields of the armies. The chains which held the warring nations to their task were not destined to be severed by military genius; no sufficient preponderance of force was at the disposal of either side; no practical method of decisive offensive had been discovered. (Winston Churchill, *The World Crisis* 1916–18, Part I, p. 19.)

This helplessness of the art of war, we can see now, was due to the slowness with which armies changed away from an old conception of warfare to a new one. All the technical means for ending 'this helplessness' were present early in the war; the petrol engine, the caterpillar tractor, the idea of an armoured vehicle capable of crossing trenches and standing machine-gun fire, the aeroplane and the light machine-gun were all available. What was not available was the idea of war as a changing art or science, affected by every change in the techniques of production and transport, and inevitably out of date if these techniques are not employed to the full. The generals who were responsible for tactics and strategy, for advice to governments and demands on industry, must take first responsibility for the failure to change when change was possible. They have a heavier responsibility: they definitely and deliberately obstructed change. But looking beyond them, and seeing their figures shrunk to the measure of reality, one sees that these generals, with rare exceptions, were the last people to be expected to welcome and develop new changes in their job. Between the time when they first learnt warfare and the time when they came to command, the job had grown so much more rapidly than they had, and had already changed so much more than they expected, that they could not be pioneers of change.

None of the General Staffs had expected that artillery would possess the overwhelming importance that I have been stressing as the characteristic feature of 1914–18. The Germans were nearest to reality in their pre-war estimates, and therefore scored in their early battles by bringing to action the huge howitzers that British and French staffs thought only fit for siege work. But the Germans had calculated on a short war and had not prepared sufficient stocks of shells for trench warfare. From all the armies, therefore, in 1915 there was a clamour for more shells, and each of the nations at war found it necessary to mobilize women for the shell-filling factories. In parliaments and in newspapers so much attention was given to 'shell shortages' that public opinion came to view the production of guns and munitions as the essential key to victory in war. The bitter war weariness of later years came largely from the fact that this key failed to open any door to victory. Artillery had become the dominant weapon but was not the decisive weapon. The difference between these two ideas requires explanation. Artillery in Napoleon's hands had been the decisive weapon, though not the dominant one, the one most used to kill enemy soldiers. He had used it to secure decision by preparing the way for his infantry columns. It no longer secured such decisions. But artillery was the dominant weapon, in 1914–18, because in those years it killed most enemy soldiers and did most damage to their defences and was the most effective weapon for hampering movement behind the front line. Some incomplete figures exist that show the change very sharply. These are official German figures, quoted by Shirlaw and Troke in *Medicine versus Invasion*:

|         | Wounds caused by infantry per cent | Wounds caused by artillery per cent |
|---------|------------------------------------|-------------------------------------|
| 1870–71 | 91·6                               | 8·4                                 |
| 1914–15 | 22·3                               | 49·29                               |
| 1916–18 | 6·0                                | 85·0                                |

Clearly there are points of difficulty about these figures.

Were no wounds caused by cavalry in 1870–71? Why are the figures for 1914–15 so incomplete? They add up to less than 72 per cent. Does it seem likely that the figures for 1914–15 can be accurate to two places of decimals, while those for the later years of that war work out at round-figure percentages? I should not like to accept such calculations if my argument depended on their accuracy. It does not. My argument is based on the main tendency they show and on a belief that the real percentages were somewhere near these figures. In the war of 1870–71 infantry did about ten times as much killing as artillery; less than fifty years later artillery was doing about ten times as much killing as infantry. That is the essential point. An immense change had occurred, and more than half that change had happened before 1914. Yet the relative importance of guns and rifles seemed to Sir John French and his staff in 1914 little different from the relative importance of guns and muskets in Wellington's time.

By 1915 the new ideas were spreading. Shell shortages, the need for long artillery preparation, 'the artillery wins the ground, the infantry occupies it'. An inevitable consequence of these (then new) ideas was that surprise no longer mattered in attack. Sir John French, the British Commander-in-Chief, wrote on 14 May 1915 to a subordinate charged with command in an offensive:

'As the element of surprise will now be absent (owing to the long artillery preparation) it is probable that your progress will not be rapid.'

Progress certainly was not rapid; and only on large-scale maps could it be discovered at all.

'In short, surprise was abandoned,' writes Captain G. C. Wynne, 'and the long bombardment and so called "war of attrition" began. The consequence was that the artillery now became the chief weapon of offence, while the infantry arm dethroned from its place as queen of the battlefield became its kitchen-maids or "moppers-up", and any method of procedure was accordingly regarded as a matter of minor importance.' (*If Germany Attacks*, by Captain G. C. Wynne, p. 52.)

Captain Wynne (who uses the word 'procedure' to mean what we now call 'battle drill' or basic tactics) has described, in the book quoted, the problem that Sir John French and other commanders had stumbled against. This was, in essence, the defensive power of well-placed machine-guns. I cannot here go into the wealth of detail which he uses to demonstrate the German development of defence by machine-gun strong-points, from the first crude rigid forms of this defence to the flexible deep defences of 1918. But the reader who wants the essence of the tactics of 1914–18 in miniature cannot do better than refer to pages 42–59 of his book.

In these pages Captain Wynne shows how two basic theories of war conflicted in 1915. The question was 'whether the infantry was to have the covering fire and support of artillery and every form of mechanized aid to help its own skilfully organized forward movement, or whether the artillery and every form of mechanized weapon were to have the infantry to "mop up" what they intended first to conquer; whether in a few words, fire-effect and movement should work simultaneously together, or whether fire-effect should be followed by movement. British G.H.Q. and French G.Q.G. adopted the latter, fire-effect followed by movement, while the German O.H.L. kept to the doctrine which all armies had previously followed, fire-effect combined with movement; and that distinction is still, to this day, the fundamental difference between the latest (1939) training manuals of the French army, on the one hand, and of the German, on the other.' (*If Germany Attacks*, p. 59.)

Fire and movement together can be decisive; fire divorced from movement can kill millions, wear out armies and states, and eventually end a war by the exhaustion of one side a little before the other side exhausts itself. The decision taken in 1915 on our side was to divorce fire from movement, killing from manoeuvre, and to hand the main business of battle over to the guns. After they had done their worst, in bombardments lasting days or weeks, the infantry would flounder forward into the resulting devastation.

But there was an alternative decision, that could have been taken. A new 'procedure' was possible, new battle tactics. This alternative belongs to the next chapter. It opens the new phase of war, the phase of the present day, which begins with the word 'infiltration' and has developed to the stage described by the word blitzkrieg. But before we start on a new chapter, with this phase of war as its theme, it is necessary to complete our study of artillery as the dominant weapon in 1914–18, and the machine-gun as the decisive weapon.

I have already, as best I can, defined my use of these terms. I use 'dominant' to mean the main arm, to the creation of which during war nations turn most of their powers, on the use of which generals and staffs concentrate their hopes and energies. A 'dominant weapon', to my mind, is that which causes most casualties to the enemy, and does most of the job of battle. But a 'decisive weapon' is more important. It achieves decision, the end of the battle, victory. It dictates changes in the shape of war. The machine-gun, not the field piece or the howitzer, governed the shape of 1914–18. It did so first by 'locking the fronts'; then in new ways it 'unlocked' them. The tank is a device for combining the fire of machine-guns (and of weapons able to root out machine-guns) with movement through machine-gun fire. Infiltration is a way of getting your own machine-guns forward through the enemy's machine-gun strong-points. As far as there was any real shape to the art of war in 1914–18, that shape was formed by Maxim's invention, together with the trench and the barbed wire entanglement. And as far as there was any real decision in that indecisive war, the machine-gun was the main factor in producing it.

# 9. Tank and Plane

In this chapter we carry the story of weapons and tactics as far as the changes obvious before the main campaigns of 1942. The whole story, as we have dealt with it so far, covers three thousand years or more from the siege of Troy to the four-year siege of Germany in 1914–18. The present chapter covers less than thirty years, not one per cent of the whole. Yet it contains so much of change, so immense a revolution in warfare, that the changes of the past seem to be dwarfed by those made obvious even in the three years 1939–42; while other and possibly even greater changes loom up ahead of us, their outlines seen vaguely through the fog of war and the thicker fog of censorship.

I summarize the main changes, and only the main changes, of the years between 1917 and 1942 as: the development of deep infiltration as a basic tactical principle, and the development of new weapons that, used together, become for a period decisive – the essential feature of these weapons, the tank and the plane, being that they can move very much more rapidly than any decisive weapons hitherto possessed by armies. With these changes attack, which had been far more costly than defence, becomes cheaper in lives and material than defence.

Decisive strategic manoeuvre had for three thousand years before 1917 been tied to the pace of men marching. Very occasionally this pace had been doubled or trebled to that of men riding fast. During 1914–18 many movements of strategic importance were made by rail. But these movements were all behind 'locked' fronts. Not until tanks and planes were fully developed, and used together in a way suited to their capacities, was the average speed of decisive manoeuvre capable of tactically 'unlocking' the enemy front raised from the three miles an hour of the marching man to the ten, twenty or thirty miles an hour of the petrol vehicle.

In an introductory chapter we singled out mobility, hitting-power and protection as the most important factors (with morale) in battle. Mobility in these brief thirty years changes from feet to wheels or wings. Hitting-power changes less; but with the development of the bomber at one end of the scale, and of the sub-machine-gun at the other, the weapons of today have hundreds of times increased the range of hitting-power and several times increased its concentration at close quarters. Meanwhile protection has undergone a change more radical than either of the other two; it has changed not merely by quantity becoming quality but by an essential change in its nature. Protection is no longer given by earthworks; it is mainly given by armour and by invisibility.

The tank was at first a weapon that *restored* mobility, but did not greatly *increase* mobility either tactically or strategic-ally. The first tanks travelled at three or four miles an hour, and like the aeroplanes of 1914–18 their range was very limited. They were thought of, by those who first proposed them and brought them into being against the vigorous opposition of official military opinion, as moving machine-gun platforms sufficiently armoured to protect their crews from machine-gun fire, and sufficiently mobile to go ahead of infantry and clear a way for them. This remained their principal function, even in the views of their most devoted admirers, until at the battle of Cambrai, late in 1917, they showed that they were inherently capable of offensive movement so rapid as to leave the in-fantry far behind them. In other words they showed themselves capable of a decisive operation: the breach of the enemy's position. The function that had been performed long ago by the armoured phalanx of foot-soldiers, and later by the armoured 'battle' of knights, was after 1917 possible, achievable, by the modern armoured force. This function was not in fact carried out fully, by an armoured force, during the next twenty years or more, until the Battle of Aragon in 1938. But in 1918

French and British tanks showed, even more successfully than at Cambrai, that they possessed the power to achieve a breakthrough, and that the question of developing this breakthrough into decisive manoeuvre was a question of how to support the tanks, how to speed up the available support for them. In 1918 tanks won a great war: in 1938 they revolutionized warfare, because used in a new way.

The tank itself grew up rapidly; and at first all its development was towards greater speed. All the French tanks of 1917, and half the British, carried machine-guns only as their weapons. Their tactical value was that they added to this decisive weapon, the machine-gun, the power to move under fire. Half of the British tanks, however, carried two six-pounder guns each. They carried these particular guns because these were naval guns which the Admiralty found it possible to spare; the War Office did not find it possible to spare, or to make, any such armament for tanks. In fact the War Office attitude to tanks was mainly confined to cancelling the orders given to construct them, whittling down the construction programmes when these were forced through by Cabinet Ministers, and staffing the Tank Corps with officers who had in some way gained a reputation for 'difficulty'. Luckily this type of officer was, under the social conditions then reigning in the British army, often the best available for a new arm developing new tactics. And it is remarkable how soon the officers of the Tank Corps developed a clear sense of the possibilities of their arm. For this much of the credit is due to Major-General J. F. C. Fuller, whose autobiography, *Memoirs of an Unconventional Soldier*, contains a blistering description of the military conservatism which delayed or prevented correct use of tanks.

By 1918 'whippet' tanks were in action, capable of four or five times the speed of the first models. These tanks showed their capacity on 8 August 1918, by penetrating up to and beyond the enemy's artillery positions and divisional corps headquarters. In other words this weapon was now capable of piercing the whole depth of a normal defensive position; what it could not do was take with it the infantry and artillery to hold the ground won. The problem of support and occupation remained unsolved in 1918; it was first solved in 1938 during the Spanish War, when aeroplanes as artillery support for tanks (and particularly the first dive-bombers used in action), and lorry-borne infantry for occupation of ground won, supplied the two necessary ingredients to make the armoured division the decisive unit in a period of warfare.

The aeroplane developed during 1914–18 more rapidly than any other weapon. The first large bomber ordered by the British Admiralty in 1914 was intended to have a speed of 72 m.p.h. and to carry six 112-lb. bombs. It was powered by two engines, each of which developed 255 h.p. This was considered a 'giant' in those days; my own first flight was taken in a Flying Corps reconnaissance machine of 70 h.p., without armament and only capable of lifting a bomb if the passenger seat was empty. Normal armament on the 90 h.p. machines used by the first squadron to which I was posted in France in 1916, was a single machine-gun which could only fire towards the rear. These observation planes mainly worked for the massed artillery of those days, observing shell bursts and correcting range. Air bombardment on the battlefield or close behind it was of practically no importance; occasional longer raids on railways, road junctions or dumps were only of importance because they slowly forced on the armies the need to move at night or during bad flying conditions.

By the end of that war the average speed of fighter planes had nearly doubled, though heavy bombers were still fantastically slow. In 1918 a four-engined bomber, the Handley-Page V. 1500, was available to bomb Berlin. Its weight loaded was 30,000 lb., as compared with 8,000 lb. for the 1914 'giant' mentioned; it carried a number of machine-guns for its own protection; it could have carried 1,000 lb. of bombs to Berlin. Equivalent development had taken place in fighters; and special types of machines had been designed to intervene directly on the battlefield either by bombing or by ground strafing. But the aeroplane during that war, though its development was rapid, was only in its infancy. At no time did any air force have one-tenth the hitting-power of the artillery of its army.

Twenty years later, during the war in Spain, air power had become adult, and General Franco's bombers normally dropped as many thousand tons of bombs on a battlefield as his artillery did of shells. The figures for the operations which, put together, constitute General Franco's counter-offensive on the Ebro in the autumn of 1938 are 9,000 tons of artillery projectiles and over 8,000 tons of bombs. It is probable that subsequent German campaigns in Poland and France would also show a rough equality between these two forms of bombardment. By 1938, therefore, air power had become capable of taking over a large share of the functions of artillery. In particular it would take over the function of supporting tanks and motorized infantry. As a flying artillery, planes could do what guns used to do in Napoleon's hands: blast open a breach in the enemy's position for decisive manoeuvre. From 1938 it was clear that the dive-bomber was the best type of plane for this decisive function.

Meanwhile the third factor necessary for the modern armoured unit was developing through the normal civilian development of road transport. A lorry in 1916 was a clumsy machine put together by hand and not designed for large-scale manufacture. By 1936, thanks to Mr Ford and those who imitated or improved on him, lorries could be turned out of vast factories, in which they flowed together on conveyor lines, by the hundred thousand. Whole divisions or army corps could move at the same speed as the tanks, only jumping down from their vehicles to go into action.

The speed of tactical manoeuvre thus in a few years leapt from 3 m.p.h. or less to a possible 30 m.p.h. or more. Normal speeds were 10 to 12 m.p.h. But the speed of strategic manoeuvre did not increase to the same extent. For various reasons – difficulties of road congestion, need to refuel vehicles, and need to harbour tanks at night are among these reasons – the armies on wheels and tracks that the Germans poured across Europe between 1939 and the end of 1941 did not often move in decisive force at a speed greater than 45 to 60 miles per day. Perhaps they did not need to move faster; the armies opposing them could be pierced, severed into fragments, and shattered by strategic moves that went at this pace. These armies all depended, in defence, ultimately on their power to move infantry on the roads. Like the Germans, these armies possessed lorries; like the Germans again they had also ordinary infantry divisions, without much transport, that moved by marching. The essential fact was that the Germans were on the offensive; these armies on the defensive. For strategic offensive manoeuvre against some part of the enemy's position, and towards key points behind that position, the Germans could use their fast-moving troops; for the shifting of troops not attacked, and reserves, towards the point of battle the opponents of the Germans had to move troops by foot as well as by lorry. They had not recognized the dominance, in that period of warfare, of the armoured and motorized 'combat team'. Because of this their armies had not so organized their forces and their transport as to make possible a counter to German blows, a counter-blow struck with great tactical rapidity and with a strategic rapidity higher than that possible for forces marching on foot. Men have marched, under special conditions, sixty miles within twenty-four hours. But they cannot often do so; and their value as fighting units is decreased by exhaustion and lack of sleep. Normally, for marching armies, strategic manoeuvre at the rate of sixteen miles a day is fairly good going. The Germans were going over twice this pace, and therefore were able to divide and get behind their antagonists, surrounding and trapping portions of their armies.

This is the story of the Polish campaign, the fall of France, the campaigns in Greece and Libya during 1941, and the opening stages of the Russian campaign in the autumn of that year. This is the blitzkrieg. It is a system of weapons and tactics capable of piercing an enemy's position (if defended by troops trained mainly according to the methods of 1914–18, or the British and French methods of 1939) and destroying the enemy army by encircling some part of it; by the speed and vigour of the break-through this system of tactics tears out a flank, or several flanks, in the enemy's continuous line, and then attacks severed portions of that line from the flank or rear.

The development of the new weapons that, through their speed of tactical and strategic manoeuvre, became of decisive importance is one half of the blitzkrieg; we have already outlined this half. The other half is the development of new tactics, the tactics of infiltration raised to a new level and a new speed.

The first idea of modern infiltration we can trace occurred in the mind of a French infantry captain, André Laffargue, whose company was attacking (with of course many other French troops on each side of it) towards Vimy Ridge on 9 May 1915. In spite of heavy losses his company broke through the German trenches and captured a ruined village beyond them. A long bare slope ahead of them, rising to the crest of Vimy Ridge, seemed empty of the enemy. Then two German machine-guns opened up, from a 'nest' which covered this slope. They were the only opposition, but in that area they held up the whole attack. For four hours Laffargue's company and another French company lay waiting for the German machine-guns to be cleared up. Then German reinforcements arrived; next day a fresh French battalion tried to advance and was stopped short, mainly by the same two German machine-guns but partly by the German reinforcements.

Captain Laffargue was intelligent; he was also exasperated. He had tried to get his own artillery, during those four hours, to drop shells on the German 'nest'. And he had realized that this artillery, at least a couple of miles behind him, could not find the target. In the pamphlet he wrote on new methods of attack he asked for heavy mortars right up with the front line of the attacking infantry; and he also asked for a new organization, a new basic tactic, for part of that infantry.

Infantry in the old days of the musket had always moved forward in line; it was still moving forward in line. (I know places where it still does.) During the development of skirmishers in the Napoleonic period the front line had become less regular and more widely spaced; later all infantry became skirmishers, and became still more widely spaced. But it still kept to a line, and the fire from this line was mainly directed straight ahead or nearly straight ahead. Laffargue suggested that in front of the normal line of infantry two special groups of men should go from each platoon. They should be heavily armed with light automatics. And he used the word 'infiltration' to describe the way in which these special groups should proceed.

The normal way was then as follows: by the time the first wave of infantry in line had reached part of the enemy trenches, much of this attacking line had been destroyed by fire, and many parts of the enemy trenches could not be directly assaulted from the front; the first attackers would therefore spread out sideways along the trenches to roll up the remaining German defenders in those trenches. This was slow and costly; the enemy knew the shape and exact position of his own trenches; his artillery, machine-guns, and defending infantry emerging from dug-outs, had great advantages over the attackers slowly bombing their way along the trench. Laffargue proposed that the two special groups ahead of each platoon attacking should press forward further into the enemy position, through any gap they found, through communication trenches or dips in the ground or any available cover, until they were in a position to take the German machine-gun nests from the flank or from the rear. They would move forward; they would fire not forward, but to the side or even to some extent behind them. The German machine-guns were protected by steel plates and concrete or sandbag cover from frontal fire; they were often at this stage in the war unprotected from the rear. And they were hidden from in front, but far more visible from the rear.

The new combination of movement and fire proposed by this French officer was directed towards countering the decisive defensive weapon, the machine-gun well posted. Unluckily for us his ideas had no effect on the French army, and his pamphlet was not even translated into English. But a copy was captured by the Germans, and it was found by them to be

a concise expression of a doctrine which exactly corresponded to the course they themselves had been trying to follow by cumbersome and slow degrees. The pamphlet was at once translated into German and issued as an official German training manual, eventually becoming the basis of General Ludendorff's text-book for 'the attack in position warfare'. It was with an elaboration of Captain Laffargue's doctrine of infiltration that the Germans so effectively broke through the British position in March 1918 and the Chemin des Dames position in the following May; and his ideas have remained the foundation of the German training manual for attack to this day. (*If Germany Attacks*, Captain G. C. Wynne, p. 58.)

The development given to this doctrine was similar to the development of the idea of skirmishing more than a hundred years earlier. At first, in accordance with the normal process of grafting new ideas on to old ones, the infantry groups 'filtering' forward were small in proportion to the main body of the infantry attacking in line behind them. Like the skirmishers of the Napoleonic armies, they were only a tenth or a quarter of the troops available. Later they became a more considerable proportion, until the great campaigns of 1918 all German front-line troops were infiltrating, and only the reserves coming behind them were expected to attempt frontal assaults in some sort of line, to mop up centres of resistance.

For these new tactics, new weapons developed. The type of machine-gun suitable for the old tactics was too heavy for the new. It was a solid piece of machinery that needed two or three men to carry it, had a water jacket to cool the barrel during continuous fire, and was used during a normal attack from positions well behind the front line of troops, or at a distance to the flank, to give them covering fire over their heads or across their front. The new tactics needed a light air-cooled weapon that could be carried by one man, to be the spearhead of the group filtering forward. And so in the middle of the war of 1914–18 we got the development of the machine-rifle or the light machine-gun, such as the French Chauchat or the German Bergmann.

An immense change in training was also necessary; that change has been spread over so long a period, in the British army, that its full effects are only today becoming visible. The basis of the new training is a new form of drill. The old forms of recruits' or barrack-square drill had been retained by military conservatism in all armies, from the days when troops armed with muskets had to be trained to repel cavalry by volleys simultaneously fired, to move in close order shoulder to shoulder without breaking rank, and to maintain under all circumstances a straight alignment 'dressed by the right'. This drill became very bad training for troops whose tactics were infiltration. They no longer learned, on the barrack-square, anything parallel or comparable to their movements during battle. Worse than that, they were conditioned to the opposite of what they should do in battle; they were made to feel normal when doing the things that would be fatal to tactical success and to their lives. They were drilled to maintain the straight line, for all too long after their only hopes of success depended on their advancing in irregular small columns or arrow-head formations moving from one patch of cover to another. It was therefore not surprising that new recruits, who had received little drill, often proved themselves more capable of developing these tactics than other units to whom the old tactics had become automatic through drill. This became particularly clear after 1918 when the new Red armies of the Soviet Union developed infiltration to its highest possible level, that of guerrilla or partisan fighting.

The idea of infiltration spread from minor tactics to grand tactics and to strategy. Ludendorff was not able to make this change; his great attacks of 1918 were tactical successes largely because they were carried out by the methods of infiltration; but they were not strategic successes because he never achieved such a break-through at such a speed that his forces could penetrate to the rear of the opposing armies, separate them and roll up the separated portions. It was not until the latter half of the Spanish Civil War that infiltration on this scale was worked out; and by that time the essential weapon permitting this application of the basic idea had been almost fully developed. This weapon, the modern tank, was then combined with swiftly concentrated aerial artillery and swiftly moved lorry-borne infantry to produce a new type of force, capable of the strategic infiltration that is the essence and secret of the shape of the war up to the end of 1941.

Infiltration in attack at first meant the movement forward of very small bodies of infantry, whose duty it was to work their way through the enemy position and gain objectives from which they could outflank those positions or take them from the rear. During this period it was still possible, and may have been convenient, to set limited objectives for each stage in the attack. Each group attacking should only go so far forward; if they went further forward they would get out of touch with those following them. But when this form of attack changed and became strategic infiltration, when it was carried out by vehicles rather than by troops moving on foot, the idea of limited objectives became obsolete. In the autumn of 1941 I was following large-scale British army manoeuvres, towards the end of which a 'German' invading force was being thrown back; unfortunately the successful commander set limited objectives for the advance of his troops, and held the more rapidly advancing units back until other units right and left of them were able to get up roughly in line with them. The result was that the more advanced of his units were unable to pierce the 'enemy' formations; indeed the 'enemy' organized his withdrawal so well that within a few hours almost all his troops were thirty miles away from our 'advancing units', most of which were not advancing at all but were waiting for orders permitting them to advance. Strategic infiltration makes necessary the unlimited objective: the foremost attacking troops have to get as far forward, as far round the flank and rear of the enemy's forces, as they possibly can. They must not check to secure their own flanks; their speed of movement, the disturbance they cause by crashing through the enemy's supply lines and command centres, will serve them instead of flank protection; or planes over their heads, and other troops of their own army coming up through the gap they have made, will protect the edges of the breach. The whole idea, therefore, of an attack by stages, from one green line to the next yellow line, becomes out of date in this new shape of war.

So does the idea of a wide front for the attack. If you are going through the enemy's wall in linear formation, each man shoulder to shoulder or a few yards apart, it is necessary to blow a wide hole in this wall. If your aim is to creep and wriggle behind that wall, you need only a 'mouse-hole' to start with. And in any defensive position there is always a mouse hole; no such defensive position can be of equal strength all along hundreds of miles of countryside.

During the civil war in Spain the German army carried out careful experiments on the question of the minimum width of attack necessary in order to breach defended lines of trenches. They found that in close country it was practically possible to start with a frontage of attack only about a thousand yards wide. But to be sure of results the necessary frontage was about two kilometres. This became their standard frontage for the blitz attack; and normally they attacked on two or three such frontages with gaps of about ten miles between them. This was their pattern at Sedan in May 1940, and also apparently in Libya in May 1942.

The best analysis of this pattern is that given in *Blitzkrieg* by F. O. Miksche, a Czechoslovak officer who was a major on the General Staff of the Spanish Republican army. His book shows that the same basic tactical ideas govern the handling in battle of a secton of an army corps; the German pattern of combined fire and movement gives to infiltration in the attack so brusque and rapid a character, when it is carried out on a large scale, that it becomes not a normal break-through (that can perhaps be checked and confined by reserves and the formation of new positions) but an 'irruption', a break-through that is also the breaking up and severing of the enemy. And he points out that the repeated success of this operation, in Poland and France,

Greece and Libya, depended largely on extreme concentration of force against the narrow sectors chosen for attack. This extreme concentration was made possible by the petrol engine. Motorization gives the blitz attack the power to penetrate almost any linear defences, because it gives the power to concentrate the force of five divisions on a mile or two of front.

But another form of defence, a non-linear form, is possible. That form Miksche, in the book mentioned, describes in principle and in theory. It is the form known as 'web defence', of which the basis is the holding not of consecutive lines but of islands of resistance capable of all-round defence, capable of continued fighting for long periods after they have been surrounded.

After Miksche's book had been written, and while it was being translated, the Nazi attack on the Soviet Union began. It began with an enormous double blitzkrieg, in which the Germans drove two 'wedges' through and round the main Soviet armies on the Polish border, and joined these 'wedges' to form what they call a *kessel*, or basin (pocket, bulge, or cauldron), in which the Russian advanced forces were trapped. This battle of Bialystok was, from the German point of view, a most successful operation, and according to their own figures was as large as the whole of the battle of Flanders (i.e. all the fighting of May 1940 up to and including the evacuation from Dunkirk). But one aspect of it must have made the German commanders somewhat worried about the future. Russian infantry isolated in the old fortress of Brest-Litovsk continued to resist even when isolated, and according to a German account (quoted by the U.S. *Infantry Journal* for April 1942), it was necessary for the Germans to leave behind a whole infantry division to contain in this citadel 'several thousand Russian troops'. The larger Russian forces cut off round Bialystok, when the two German wedges met behind them, also continued to resist. Though surrounded they did not surrender. And it is clear that weeks after the battle began, and even weeks after German official announcements had claimed that it was successfully ended, Russian 'islands of resistance' remained in the more difficult and marshy parts of the area. During the next three months, in fact, some of the Russian units forming these islands of resistance seem to have been fighting their way towards the Soviet lines; others were splitting up into guerrilla detachments to harry the German communications.

Although the defensive doctrine of the Red Army had not stated, or had not made clear, the idea of 'web defence', their troops began to put this idea into action as soon as the fighting started. Part of the reason for this must have been the experiences of the Russian Civil War, in which the rapidly moving fronts and partisan and guerrilla fighting crystallized around positions defended in isolation, and particularly around towns, villages and other road centres. Part of the reason for this development must also have been the study made in Soviet officers' schools of the development of German defence tactics during 1915–18. This German system of defence by strong-point is the main subject of Captain Wynne's book *If Germany Attacks*, already quoted. The staffs of the Red Army must also have realized the development of defensive systems during the Spanish Civil War. This development recapitulated the development of 1914–18, and then went beyond it to a much greater depth of defence and much more reliance on the fortified strong-point.

It is sometimes forgotten now how far even the British army's defensive system in 1918 had gone in this direction. The official *Military History of the War* for 1918, compiled by Sir James E. Edmonds, states on page 257 that 'there were continuous lines of trenches in the Forward Zone, but the garrisons as a rule were disposed not in lines but in posts, strong-points, etc., for all-round defence'. The reason why this system failed in March 1918 may be gathered from the following quotation from the same page of our official history:

There was a general objection among the fighting officers to the distribution of the troops in small packets, the 'blob' system of defence, as it had been called, in derision, before the War, for it was not a new theory. Some even went so far as to call the policy 'suicidal'; for without strong reserves to counter-attack the enemy if he penetrated the intervals, the surrender or annihilation of the posts must only be a matter of time. The majority of experienced fighters, in view of the inadequate number of men available and lack of strong counter-attacking forces, would have preferred a definite line of resistance in each zone, with posts, machine-gun nests, and switches, arranged in depth behind it to limit any enemy entry into the line.

The British soldier has times without number defended isolated posts to the death; and he did so on the 21st March 1918, and was to do so repeatedly during the next week. But he prefers to fight in line. An old N.C.O. of 1914 summed up the new system in discussion with an officer: 'It don't suit us. The British Army fights in line and won't do any good in these bird cages. . . .'

There were too many inexperienced young officers and too many untrained young soldiers to ensure a reliable garrison for every post, even without the special trial to which the fog subjected them. The platoon commanders were unable to exercise control over more than the posts in which they had elected to be, the section posts were unaware whether those on the flank were holding out or had been captured, with the result that there was a lack of confidence on the part of small and, on account of the weather conditions, isolated garrisons. To British troops, whose instinct is to fight it out where they stand, there came no thought of 'elastic yielding', and considerable doubt existed as to whether the garrisons, when the enemy was already in rear of them, should hold on to the last regardless of what was happening on the right or left. Some even of the best of the new officers did not realize that they must use discretion as being 'the man on the spot', and that even orders to hold on may in extreme circumstances be disregarded. No warning seems to have been given any brigade or battalion commanders, and therefore none to the lower ranks, that in certain circumstances there might be an ordered retreat; divisional routes had been reconnoitered for this, but information of such a nature was certainly withheld from regimental officers.

It is clear from this quotation that the training of these British troops had not conditioned them to the new way of fighting. They had been drilled and exercised only to fight in line, whether attacking or defending. They had not realized that the German system of defence by strong-points, so devastatingly effective against British attack in the Previous years, implied a retraining of the army in which men were not only taught new tactics but also taught the reasons for them; or that it implied the utmost initiative by junior commanders and their readiness to move in any direction required by the course of the battle. The official history refers to the 'instinct' of British soldiers to fight where they stood. It is very difficult to consider this a scientific use of the term 'instinct'. It is probable that the author means that the men had been conditioned to certain actions and attitudes of mind, and it is tragic to reflect that the same process of conditioning men, to what amounts in modern battle to stupid passivity, still continues in many parts of the British army and even in the Home Guard.

General Ludendorff had realized the new conditions of defence by the winter of 1916 and in his *War Memories* writes of 'a more active defence ... it was, of course, intended that the position should remain in our hands at the end of the battle, but the infantryman need no longer say to himself, "Here I must stand or fall," but had, on the contrary, the right, within certain limits, to retire in any direction before strong enemy fire.'

Here then we have the remarkable fact that General Ludendorff, explaining the basic idea of successful German defence, and the British official history explaining the failure of a British adaptation of this system of defence, both refer strongly and clearly to the need for retreat to be considered, when necessary, as a normal part of tactics. And no unprejudiced observer looking at the history of British arms, during that part of the present war occurring before this book was finished, would think it probable that the training of our army in the summer of 1942 should normally consist of 80 per cent training in attack, 10 per cent training in the holding of positions, a few per cent of orderly withdrawal covered by a rearguard, and only in some units and in a tiny proportion, the teaching of why and when to retire, and why and when to hold on to isolated or surrounded positions. In most units the idea of withdrawal is treated as something shameful, and the idea of being cut off as something desperately dangerous. Yet the Russian defence of 1941 was necessarily based on both these ideas; and as the Russians grew accustomed to the conditions of modern fighting and developed the theory and practice of a defensive system approximating to 'web defence' it became increasingly clear that each blitz offensive by the Germans was less of an actual victory and more of a stalemate.

Then the Germans had to turn to the defensive in the Russian winter. And it soon became clear that their defensive system was based entirely on the idea of strong centres of resistance covering the railway lines and other transport facilities and capable of standing siege. One of these centres of resistance at Rzhev, and another at Staraya Russa, seem to have been partially or completely cut off from supplies (except by air) for most of the winter of 1941–2.

The Germans, boasting of their success after the winter was over, described these besieged centres as 'hedgehogs'. The word has a curious history, as a military term for defended localities. When working with F. O. Miksche on the book *Blitzkrieg* already mentioned, I published in May 1941 a popular summary of the idea of web defence in an illustrated weekly, in which I described islands of resistance organized for all-round defence as 'bristling with arms as a hedgehog bristles with spikes'. And I wrote of the men manning certain linear posts towards the rear of a defensive system as being able, when necessary, to 'roll themselves up into new hedgehog islands'. This article was reprinted in part in a German illustrated weekly. (The general line of the German comment was: what fools these English are – look, they are only just finding out these things which we knew long ago.) Apparently the Germans like the simile of the hedgehog, and applied it during the winter to their own islands of resistance in Russia. Then General Rommel began the battle of May–June 1942 in Libya; and one of the first dispatches to reach the British press, describing the defensive tactics of British and Free French troops in this battle, referred to our positions as 'hedgehogs' round which the German units flooded before being forced back. This battle is continuing as these pages are written; its first stages have gone against us; its outcome cannot be foreseen; but the fact that we are relying for our defences partly on 'islands of resistance' is clear for the first time in this war. We shall hear more of 'hedgehogs'. They are the Roman camps or strong-walled castles of today.

One of the essential features in any modern system of defence, and particularly in web defence, is the use of an immense number of land mines. It is clear from descriptions of the Russian campaign that both sides used them in quantities never before employed. And it is also clear that the British defences in Libya, like General Rommel's defences south of the Bay of Sirte in an earlier phase of the campaign, now consist largely of vast minefields. In fact the early part of Rommel's attack in May 1942 consisted first of his opening two breaches through our minefields, then of his armoured forces forming a *kessel* or cauldron, round and behind the ten miles of minefield between the two gaps, and then punching out from the rear the troops holding these ten miles and thus widening his gap. When the main bulk of his armoured forces had flooded round the southern flank or through the original narrow gaps, and were turning back to take our defending troops between them from the rear, we got the usual optimistic Cairo stories that Rommel's armoured forces were retreating. It seems almost incredible, after so many examples have been given to the world of the normal German tactics of the *aufrollen*, the rolling out that follows the thrust, that even the Cairo spokesman should not recognize at this late date the pattern of a blitz offensive. But strange things happen in the Middle East; strangely enough, for example, the illustrated weekly in which alone the 'hedgehog' system of defence and the large-scale minefield had been advocated – I can trace no similar advocacy in any other British periodicals – was partially banned from the Middle East by government order not long ago. This ban was not imposed in time to prevent the idea of web defence soaking into troops in Africa; but the official dislike of critical thinking, of which this ban is a tiny example, must be partly responsible for the fact that our forces in Libya, in May and June 1942, were only partially organized for modern defence and were deployed mainly in a linear defence without depth.

It take thorough retraining, with careful explanation, to make troops fight in a new way. Our failure in March 1918 was due, as I have pointed out, to lack of this retraining; to quote another example from the *Official History of the War*:

'The enemy method of firing machine-guns in enfilade rather than straight to the front led the partially trained troops to imagine that they were being fired on from behind or out-flanked, and they often retired for this reason alone.' (1918 volume, p. 401.)

Partially trained troops to whom the new tactics have not been properly explained will always be made uneasy, if not useless, by enemy infiltration. How much of our failure in June 1942 in Libya was due to this factor cannot yet be judged at the time of writing. But clearly the lack of modern tactical theory and of retraining on that theory was largely responsible for earlier disasters, for Singapore and Burma for example.

The pattern of blitz warfare includes other elements that will be noted in the next chapter, in which I attempt to analyse the probable future tendencies in warfare. They need to be stated here, briefly, to complete our outline of the changes in weapons and tactics up to the full development of the blitz-krieg. The first, and to the popular mind the most striking, is the development of airborne troops – parachutists and troops landed from carrier planes and gliders. Such troops played a minor part, but one of importance, in the Battle of Flanders. They have only once, at this writing played a decisive role in battle; they took Crete. They will have more importance in the future.

The second element is the development of armed civilians, as Home Guards or guerrillas, and of guerrilla warfare on such a scale as has not been seen since Napoleon's day. The importance of this aspect of modern tactics is considerable; I argue in the next chapter that it will become the decisive factor in war.

Then a third element in the pattern of blitz warfare: the shaping of protective weapons, anti-tank and anti-aircraft, with which to answer and check the new dominant tank–plane combination. These protective weapons have a history of their own; they begin as things separated, organizationally and tactically, from the normal weapons of infantry. They end as infantry weapons, closely and organically connected with the men who, without them, can be cancelled from the battlefield. Among these weapons some show a significant tendency towards a multiplication of functions. Thus the German 88-mm. gun is first an anti-aircraft weapon, then becomes in addition a field gun, and then becomes the most powerful anti-tank gun in normal use. It changed in this way from a single-purpose weapon to a triple-purpose weapon during a year of fighting in Spain. And the whole idea or principle of several purposes or functions in a weapon is so alien to our military conservatism that this gun is a 'surprise' to us in Libya in 1942, five years after it had first been used in these three ways.

Infantry, in this period of the blitzkrieg, is an arm that fights tanks and planes as well as men. It can only do so if it is given new weapons: explosives, anti-tank mines and grenades, anti-aircraft and anti-tank guns. It is at the same time given field guns, directly under the control of the infantry or regimental commanders, because owing to the rapidity of movement of modern battle there is no longer time for separate arms in separate organizations to function together. In this way an infantry brigade or regiment becomes a unit of all arms, and even smaller units become self-contained 'little armies on their own'. This process develops in the direction indicated by the words 'combat team'; any part of a fighting force at any time tends to become a team of several arms closely integrated together.

On the same principle air, sea and land forces are integrated by the Germans in what they call their *Wehrmacht*, a single fighting force under a single command and a single staff. Each of the smaller units that make up this fighting force, whether they contain ships and planes, or ships and troops, or all three together, is given a single commander; and the three arms are further linked (as within ground forces each part of the combat team is linked to the others) by radio. This is used in action without the slow and painful business of coding and decoding still imposed on our forces by an out-of-date idea of security; the Germans sending messages *en clair* protect their security by speed of action, rather than by hampering attempts at secrecy.

These then are some of the lesser developments of the blitz-krieg pattern of warfare, the pattern that has dominated the present war until 1942. But in that pattern we should not allow these smaller threads to obscure the main design. The main thing in the design of modern war, from the Aragon Battle of 1938 to the Libyan fighting of 1942, is a revival of the armoured phalanx, of the concentrated column hitting hard on a narrow front, armour on tracks and wheels playing the same role as the armoured foot-soldier of Alexander or of Caesar or the armoured chivalry of William the Conqueror. This armoured phalanx was handled in a new way. Because it provides a dangerous target when massed closely together, a target for enemy bombers or long-range artillery, it can only be massed when needed for decisive action, for the break-through. Over and over again I have heard the complaint, from those fighting in Libya and the Western Desert, that our tanks were scattered when General Rommel's tanks were concentrated. Most of those who complained stated the lesson of these campaigns to be the concentration of armour. Yet on 13 June 1942, the decisive action in the last of these campaigns to take place before this book was finished, our own concentrated armour was destroyed by enemy gunfire. And occasionally we have heard of General Rommel's tanks being so scattered about the desert that our own forces could not find them and bring them to action. It is clear that German tank tactics include alternating concentration and dispersion.

The concentrate for the *schwerpunkt*, the rolling thrust

that zig-zags through an enemy dispersed in defensive positions; having achieved their break-through, they fan out in the *aufrollen* to find the next weak point, to take defensive positions from the rear and to exploit their penetration to the full. Faced by new resistance they concentrate again to by-pass it or break through it, and then fan out again. And this alternating concentration and dispersion is the essential tactical pattern not merely of the Panzer divisions but of all German combat teams however formed.

This pattern of tactics has the same essential shape on the map as the movement of Napoleon's army corps and divisions during a campaign. These forces marched separately but struck together. Between battles they spread out not merely to have room in which to live and move (and live off the country) but in order to overlap and threaten from different angles the enemy's positions. The tactics of the armoured phalanx, and of modern combinations of infantry and other arms, are not tactics derived simply from the weapons that modern industry produces or from some theory or skill of the Germans. They are tactics embodying in a modern shape all the proved expedients of the masters of warfare from Alexander's day to Napoleon. Their aim is decision by envelopment – by pressing and jamming the enemy closely together, as the legions were pressed at Cannae or the Austrians at Leuthen – and envelopment is sought by a swift irruption that divides the enemy's army. They are tactics that can be justified by reference to Epaminondas or to Clausewitz. It is surely time they were recognized as valid by those who control the British army; and recognized also by them and the British people as necessary methods that must be learnt before we can go beyond them.

## 10. Change Goes On

The essence of the thing I have been trying to do in this book is not statement of a pattern of war; it is statement of a pattern of change in war. I have not been trying to establish the thesis that this, that or the other is the essential feature of modern tactics; I have been trying to establish that modern tactics are changing, as modern weapons are changing, more rapidly than ever before; but that the lines on which they are changing parallel the lines of past change and should be to some extent predictable. And is is only if we can predict, foresee, the lines along which war is changing that we shall be able to establish a 'doctrine' of warfare, an integrated system of choice and design of weapons and retraining for new tactics, that will not merely rival the methods of our enemies but will go beyond them and be superior to them.

The first tendency in any armoured period, as we have already pointed out, is towards the creation of a force whose essential shape is that of the phalanx – a 'heavy hammer' of armour. Tanks were not used in this way when they were first invented. In 1917 or 1918 tanks went into the attack spaced out in 'waves' or lines, each machine fifty or a hundred yards away from its neighbour to right or left of it and each 'wave' followed at some distance by another wave. And this remained the typical method of tank attack in most peace-time man-oeuvres between the two great wars and in the early stages of the Spanish Civil War. In the middle of that war, after some tentative experiments near Bilbao, the Germans first tried massing 150 tanks on a narrow front to achieve a break-through in Aragon. A little later army manoeuvres in Germany showed that the idea of a phalanx of tanks had been adopted as the Nazi solution for the stalemate of trench warfare. Later still, in the Polish Corridor and at Sedan and in the Monastir Gap and on the plains of the Ukraine the Germans showed us in detail the methods of the blitz attack of which the spear-head is this phalanx of tanks. It normally included, before 1942, the massing of all medium and heavy tanks of a division – say 200 machines – in a 'combat echelon' advancing on a front five hundred to seven hundred yards wide. This frontage would be filled by perhaps twenty tanks, each about thirty yards away from its neighbours on either side, and each of these twenty machines leading a file or column containing nine others, spaced at intervals a little greater. This 'combat echelon' would cover a square of ground, each side of which would be about one-third of a mile. Its advance would be covered by lighter units and by engineers specially trained to deal with anti-tank obstacles; behind it would come lorry-borne infantry and mobile guns to mop up, and to widen the breach made in the enemy's defences. It was a combination that was irresistible – until Moscow and Leningrad resisted it.

At some point in the autumn of 1941 it became apparent that the period of the armoured force as a heavy hammer was probably ending, on the Russian front. It could still continue to be used in this way by Rommel in Africa, against a defence that neglected depth; it could only be used once or twice more in this way, effectively, against the Russian defence which was all depth, which consisted of hundreds of miles of guerrillas, combat troops, counter-attack troops, fortified towns and reserves, from front to rear. And at the same time there appeared an obvious contradiction between the type of force the Germans were using and the functions for which that force was intended. A heavy hammer does not infiltrate. We have described the 'team' or combination of tanks, planes and lorry-borne infantry as the necessary combination for long-range strategic infiltration; but this grouping of tanks into a phalanx – ideal for penetrating a linear defence – was obviously unsuitable for continued infiltration over immense areas of a web defence. The phalanx was intended for a single sharp decisive hammer blow lasting only a few hours; then the tanks should be dispersed in the *aufrollen*; to keep these tanks together in a mass when enemy bombers might find them, or enemy artillery catch them while a concentrated target, was to take enormous risks. Therefore by the time the campaigns of 1942 opened tanks were no longer used as a phalanx or spearhead on the Russian front, or were used in this way extremely rarely.* And the Panzer division of 1942 had fewer tanks in it, and more guns, infantry, planes, and pioneers.

Turn back to my second chapter and glance again at the brief summary of how fighting changed from Epaminondas to Alexander. The heavy Macedonian phalanx had to be split up into more manoeuvrable brigades and integrated with other arms, with light troops and troops using projectile weapons and all the rest. A process strikingly similar to this has been going on in Russia. And the Russians are ahead in the development of this process. It is clear that they began their war in June 1941 with a much less definite idea than the Germans about the use of the armoured phalanx, but a much more definite idea than the Germans about the need for a combination of all arms and all methods of fighting, a combination in which infantry recovers its premier place on the battlefield. And as the German massed tanks lost their impetus, and were forced to divide into smaller packets in order to carry out deeper infiltration, or to attack on wider fronts without hope of decisive irruption, the Russians gained (at a heavy cost) the necessary practical experience which gave them the power to group around their tank brigades, and group above them, the other arms and forces with which these brigades became integrated.

The tank still remains the dominant weapon, just as, long after Alexander, the armoured foot-soldier of the Roman legion remained dominant in battle. But it can only achieve its

*'A fortnight ago a surprise announcement was made to foreign journalists by the militarists of the Wilhelmstrasse: the Panzer spearhead is no more, they said in effect.' – *Evening Standard*, 24 July 1942.

ends when part of a 'combat team' of mixed arms. The typical German Armoured division of 1940 was already becoming such a combat team; it possessed its own aircraft and artillery, its own regiment or regiments of mobile infantry, its own engineer or pioneer units for the construction or destruction of anti-tank defences. But what the Russians seem to have done goes beyond this; their tank brigades and divisions are flexibly united with the infantry, and any unit consisting mainly of infantry may have as a constituent part of it either tanks or planes or both, if these are needed for its function. Commanders of such combat teams are no longer purely infantry commanders or tank commanders; they are in the old sense of the words 'general officers', that is to say officers taking a general command of all the arms and troops engaged in an action. No other sort of general is possible today.

Tanks were tragically mishandled during the period 1917–38 when they were tied closely to infantry advancing in line on foot. Then for a period, the period of the classical blitzkrieg, they were released to prove their power to achieve the breakthrough against linear defence. But then they came up against defences, in Russia, that they could not break through; these defences could be said to extend throughout the depth of the whole country. No 'irruption' could get beyond them; there was no beyond. In order to muscle and wrestle further and further into this web – like a man wading through an underbrush of flypapers – each armoured force had to become a complete 'army on its own'; and each of these combat teams had to split down into smaller teams for mopping up here, for further penetration there, for seeking a weak point in this or that direction. The forces infiltrating were being themselves 'filtered'. And the defensive filter was designed with a special aim in view: to hold back the enemy infantry and separate them from the armoured forces. This indeed is the main aim of web defence – to cut the armoured spearhead off from the body of the spear that follows it. And because this is the aim, forces attacking against web defence have to carry to the furthest limits possible the process we have been describing, the integration of infantry and armour. This integration is tactically carried to its limit when tanks attacking form a hollow square within which lorried infantry move forward.

The development of warfare in the past thirty years has greatly altered the traditional relation of infantry to other arms. In 1915 and 1916 infantry ceased to be 'the queen of the battlefield' and the theory was developed that 'artillery conquers the ground, and the infantry occupies it'. Then in 1938–9 the new theory and practice was developed by which the tanks conquer the ground and the infantry occupies it. But already this is changing. Though there are more tanks and planes than ever, the fighting value of infantry in its new form, as main constituent and basis for combat teams of all arms, is steadily increasing. I believe it will increase further, that infantry will be again the decisive arm. But it will only be so by taking to itself and making part of itself the dominant arms – tanks, planes and artillery – and by moving before action and to some extent during action at the speed possible to vehicles. I do not believe that we are going back from the blitzkrieg to Verdun; manoeuvre by vehicle will continue to be the decisive form of manoeuvre. But in manoeuvre by vehicle the emphasis will change from manoeuvre by tanks backed by infantry towards manoeuvre by mechanized infantry protected by tanks. The change may not seem a great one. But in every practical aspect of training for attack and defence, and of planning and carrying out these operations, this change of emphasis will make the difference between a real victory and a Pyrrhic one. For a period, even against web defence, massed tanks may still win their Pyrrhic victories, as the Macedonian phalanx did for a period against the Roman legion. But these victories will grow more and more costly and less and less effective. The future of battle lies with the combat teams or 'battle groups' that can combine with, and at the same time act against, armour.

Some development in this direction has been shown by the British army. During the campaign of November and December 1941 in Libya mobile columns of all arms were formed, in

a typically British and untheoretical way, on the initiative of
Brigadier Jock Campbell, v.c. They were locally known as
'Jock Columns'. Later they seem to have been termed 'Battle
Groups'. They were improvisations, and it is valuable to be
able to improvise in war. But improvisation has its limits; the
British army was neither so trained nor so organized as to
make the operation of forming and using these columns
normal, standard, and well understood by all. This is one of
the cases where to have a theory is in fact more 'practical' than
to have none.

Armour and in-fighting go together. The tactics of the
phalanx of Alexander were shock tactics. So were the tactics
of the phalanx of tanks in the blitz attack. They were
shock tactics in the sense that the phalanx of tanks tried to
get to close quarters with the enemy as rapidly as they
could; but they did not use shock weapons such as the pike or
bayonet. They used projectile weapons at relatively short
ranges. It was by the concentration of their fire on a narrow
front and at short ranges that they achieved a shock effect.
And this effect was overwhelming against troops deployed in
linear positions that could be directly approached and even
overrun by the mass of tanks.

But when troops were deployed defensively in fortified
towns and villages into which tanks could only penetrate at
great risk, and in islands of resistance, 'hedgehogs' bristling
with weapons in all directions and either naturally tank-proof
or made proof against direct tank assault by the use of deep
minefields, tanks could no longer employ shock tactics in the
same way. The pendulum began to swing back from shock
towards fire. The tanks had to use fire, and they had to be able
to use fire effectively at any range up to the limit of direct
visibility. Therefore the standard German tank, on which they
concentrated the productive capacities of Europe during 1940
and 1941, carried an artillery piece equivalent in size to a field
gun. So did the typical medium Russian and American tanks
produced in the same period. The typical British tank of the
same weight made in the same period, unfortunately, had no

heavier armament than a two-pounder anti-tank gun, throwing a shell only one-seventh the weight of that of its rivals and opponents. This tragic mistake was made partly because of production difficulties, partly because the officers and officials responsible for the design and production of tanks had no theoretical view as to their functions, or had a wrong theoretical view.

As far as there was any theoretical view of the function of tanks obvious in the British War Office and Ministry of Supply in 1940, it was the view that tanks were cavalry, to be used for reconnaissance and security, or the alternative view that they were anti-tank weapons with which to stop enemy tanks. Our generals did not think of the armoured cavalry of history, the decisive arm of the Middle Ages; they thought of the light foxhunting cavalry whose disappearance from the battlefield they so much regretted. (This light cavalry had never needed projectile weapons of any great range; the horse artillery built to accompany it was always light stuff.) They therefore asked for tanks with just enough weapons, as they thought, to protect themselves. And they actually called many of these tanks, during 1940 and later, 'cavalry tanks'. Their alternative theory was that tank should be used against tank; our armour should seek out and destroy the enemy armour. This is a completely different theory from that of the Germans. The German theory is that armour should be used against unarmoured men, and even against the weakest units and positions of these unarmoured men. The British theory was that our tanks should be used against the strongest enemy forces, their Panzer formations.

The idea that tanks should be used mainly against other tanks led, in British minds, to an astonishing misconception: British politicians and generals began to think of tanks as if they were battleships and lighter naval vessels. A fast medium tank became known as a 'cruiser' tank, and battles in the Western Desert were described as if they were naval engagements. This is a misconception because naval vessels are designed for use as floating gun platforms and the guns have only one main purpose – to sink other ships. Tanks are moving gun platforms on land, but the purpose of the guns they carry is to destroy enemy armed force. This armed force is not mainly available for destruction in the form of enemy armoured vehicles; it is mainly available for destruction in the form of unarmoured men and vehicles, and guns with little protection. This misconception is not responsible for the grotesque under-armament of our tanks; as a naval country we know that armoured vessels must carry big guns. But it is responsible, perhaps in conjunction with a misreading of Clausewitz, for the grotesque handling of our tanks, their use as if they were fleets. The purpose of a battlefleet, in classical naval theory, is to bring the enemy battle fleet to action, destroy it and then rule the seas. No such function could be carried out on land by an armoured force, as soon as tank-proof islands of resistance were formed by the enemy. The 'battle fleet of tanks' was then reduced to cruising between these islands, which it could not 'rule', and it was soon found to lose heavily in the 'narrow waters' between these islands.

The misunderstanding of Clausewitz was a simple one, already exposed by some of the more intelligent theorists of warfare. Clausewitz laid down that the aim of battle was the destruction of the principal armed forces of the enemy. But he did not mean by that that battle should consist mainly, or even in part, of an attempt to close with the main body of the enemy in a head-on encounter. The confusion here is between the aim and the method: the aim is to destroy the enemy army, but the method cannot simply be defined as 'close with the enemy army and destroy it'. The main forces of the enemy can in some cases best be destroyed by indirect approach to them, by the baited attack that lures them to fight on the ground you have chosen, or by combining resistance and counter-attack on the main battlefield with crippling diversions far from that field. Wellington and the guerrillas in Spain helped to weaken, and therefore to destroy, Napoleon's Grand Army near Moscow. It is a piece of military childishness to believe that military force can often be used as the dumb boxer uses his fists, without feint or stratagem, going straight for the enemy's main strength and hammering at it.

On the contrary, tank tactics are likely to develop in the future in such a way that attack is normally against the most vulnerable part of an enemy motorized infantry combat team, its supply columns and vehicle parks. In defence tanks will be used for the counter-blitz, the counter-attack that does not go head-on against the enemy's armoured forces but curls round behind them to catch the traffic jam of vehicles and troops coming up to their support. It is probable that the present tendency towards heavier tanks carrying heavier armament and armour will continue for a time; but a counter tendency may begin and light machines may have more importance on the battlefield if guerrilla fighting and airborne forces acquire the importance which I believe they will do. Only the lightest of tanks can be landed from the air; and the heavy tank is too greedy for fuel for it to be used as an auxiliary for guerrillas.

The blitzkrieg can be considered as reaching its full development in the two years 1940–41; during 1942 it scores full successes only against obsolete methods of defence; against the modern methods of defence shown in Russia it may still score lesser successes at a greatly increasing cost. We have not yet reached the period when armour decreases in importance in warfare – a period parallel to that following the battle of Crécy. But we have reached the period when armour used alone decreases in importance, and auxiliary arms supporting and closely integrated with the armoured units become of increasing value and essential to the success of armour. We have entered the period like that before Crécy, the period when the crossbow and longbow are essential auxiliary arms. We know that in the past when armoured forces have had to rely more and more on projectile auxiliaries, the time has come when the projectiles of these auxiliaries eventually pierce the armour and sweep it off the batlefield. Are we in sight of such a period yet?

We are not yet in sight of this. The auxiliary arms are becoming more important, but they cannot yet so certainly meet and penetrate armour. They are however beginning to be quite dangerous to armour. The three auxiliary forces that matter most today in conjunction with the tank, and form part of the essential combat team used for a blitz attack, are the air arm, mobile artillery, and mobile infantry. We cannot estimate when – if ever, in this war or in our lifetimes – one of these auxiliary arms can be developed to a level which 'negates' and cancels out the fighting value of armour. But we can see the probable line of development of each tending towards this end. This probable line of development is, for the air arm, the tank-busting plane; for artillery, the treble-purpose weapon; for infantry, tank hunting along guerrilla lines and the use of high explosive in great quantities in the form of mines and grenades.

The air arm has a very definite role to play in the theory and practice of the blitzkrieg; its role is that of a flying artillery which can be concentrated very rapidly to open a path for tanks in the attack and to support attacking infantry. This role is completely different from long-range bombardment; an air force mainly built up to bomb an enemy country does not possess the right machines or the right training to form part of a combat team gaining tactical and strategical decisions on the ground. The typical machine used by the air force of the blitzkrieg is a dive bomber. This type of machine is not, as some have stated it to be, particularly valuable because of its accuracy. Its value lies in the fact that it comes close to its target, and can therefore distinguish this target under battle conditions more easily than a high-level bomber; and secondly in the fact that it approaches its target in such a way that anti-aircraft fire against it is handicapped. The normal high-level bomber has to approach its target flying level or in a shallow dive; the anti-aircraft gunner below after estimating its height fuses his shells to burst at that height. The high-level bomber must also fly straight, in the approach to its target; the A.A. guns can therefore be trained on to its line of flight. The dive bomber on the other hand circles to spot its target and 'peels off' from its circle to approach the target from an unpredictable angle. It loses height when diving so rapidly that the A.A. gunner is at a great disadvantage. And when an A.A. barrage is being put up covering a certain area of sky, the diving machine gets through it more rapidly than the machine flying level. The dive bomber also has considerably more moral effect on ground troops until they get well used to it.

It seems almost certain for these reasons that a type of machine which normally dives towards its target will continue to be the most effective for co-operation with tanks and infantry, and for action against these ground forces. But if there is to be considerable development of planes as tank-busters, it is equally clear that the present types of dive bomber are not so efficient as the types carrying a cannon of some size, or alternatively carrying a propelled bomb which can be aimed like a cannon shell. The Russians in their efforts to stop the blitzkrieg have developed both these types. The Americans have to some extent developed the former. It is more effective against tanks because tanks are small targets, which must be hit directly to secure a knockout, and because the accuracy of a cannon shell or propelled bomb is naturally much greater than the accuracy of a bomb that drops under the influence of gravity, moved by its momentum and the winds. The Russian propelled bomb appears to be fired forward by some form of rocket, which may be a more effective method of propulsion for larger airborne projectiles than discharge from a tube.

The other development of the air arm to be expected is the use of planes, more and more, as transport rather than as fighting vehicles. The Germans at the time of writing are said to employ 16 per cent of their air force for this purpose. Their transport planes are used to carry parachutists, to land troops on captured aerodromes, to tow gliders that can land almost anywhere and to take supplies to isolated forces infiltrating forward or holding islands of resistance surrounded by the enemy. While an airborne army may prove of very great strategic value, it may turn out that an almost equal value is to be gained from supply by 'the road through the air'. Supply questions have always hampered the more adventurous generals and have often prevented them from rapid advances otherwise practicable.

Captain Hugh Slater has suggested* the idea of an airborne army consisting largely of armoured vehicles as the ultimate trend of modern mechanized war. While I agree to the extent that I have stated, that the plane is likely to become more valuable as transport than it has been in the past, I feel it likely (because of the limitations on weight that can be lifted) that the extension of function which will occur with increased air transport will be used to assist guerrilla forces rather than for the transport of heavy weapons and vehicles. In some cases the best support for guerrillas, or for lightly armed forces such as the Chinese armies, will be the transportation of heavy weapons and armoured fighting vehicles; but it is a far step from this to the transportation by air of complete armoured fighting units.

So long as the tactical pattern of the blitzkrieg lasts, the primary fighting function of the air arm will be tactical and strategical support for, or struggle against, the mobile combat teams on the ground that are capable of decision. But when the blitzkrieg is halted and its authors are forced – as they have been in Russia in 1942 – to turn towards a tactical policy that they themselves describe as *aufreiben* (grinding, slicing or grating, a policy of attrition in which they try to 'grind' through a strong web of defence) the air arm is likely to be the loosening agent by which war is made mobile again. And this development obviously will take the shape more of the transport plane and the glider than the heavy bomber.

Second in our list of auxiliaries now capable of hampering or assisting the armoured force is mobile artillery. The two developments in this field already occurring and likely to increase in importance are the placing of guns on tracks so that they can go wherever tanks can go, and the development of the treble-purpose weapon.

The Germans had developed by 1940 an 'assault gun' consisting of a 105-mm (4·5-inch) gun mounted on the chassis of

*In *War into Europe*.

an obsolete tank. The gun carried a shield protecting the gunners from enemy fire coming from the direction in which the gun was directed. But the combination was not a tank, for this armoured shield was not carried round to the sides and rear of the machine. It was not a tank, but it was artillery assimilating itself to the tank. It is to be expected that this assimilation of artillery to the tank will continue, large mortars and gun-howitzers, up to perhaps the 6-inch 60-pounder, will probably be put on tracks; others that remain on wheels will be so arranged for transportation that they are in position facing forwards rather than to the rear while being towed. This gives the possibility of catching fleeting targets over open sights in the swift movement of mechanized battle.

A certain number of large tanks may in the future carry heavier armament than the 3-inch piece of field artillery that is, at the time of writing, standard on most medium tanks in Germany, Russia and America. But it is not likely that many 'real' tanks will be built carrying much heavier armament than this weapon. The Germans have armed some of their machines with 5·9-inch mortars, but these are mortars with a restricted range. The reason why it is unlikely that tanks as such will carry much heavier armament than the 3-inch is that it is almost impossible to design armoured protection for a heavier gun, and give the crew working it sufficient space to acquire a fairly high rate of fire. Three or four men can work behind the armoured shield of the assault gun. But to fit three or four men into a single turret, around the breech of a gun of some size, means that an inordinate weight of armour must be used, and even then the men will be cramped. Also from the tactical point of view there seems some waste involved in giving an engine to each separate piece of artillery; this is particularly true when a force is turning, perhaps temporarily, from the offensive to the defensive. It is not necessary to leave much transport within islands of resistance, but it may be necessary to leave guns; therefore for certain purposes it is better that the prime mover which shifts the gun should be in the form of a tractor or 'dragon' rather than in the form of an

engine built into the gun assembly. For these reasons it is unlikely that all artillery of field and medium types will be carried
in tanks; but it is very likely that much of it will be carried on
tracks, like the 'assault gun'.

The anti-tank gun as such will, I believe, become obsolete. It
began in the period between the two wars as a light weapon
usually of 20 mm. (less than one inch). As a very rough and
approximate rule you can say that a weapon of 20 mm. has a
good chance of piercing armour 20 mm. thick; and a weapon
of 50-mm. armour 50 mm. thick. The smallest size of anti-tank
gun therefore could sometimes deal with armoured cars and
light tanks, with about an inch of armour plate round them,
but even in 1937 in Spain it was clear that this was only possible at fairly close ranges. The 37-mm. guns tried out in Spain
proved capable of piercing the armour of many of the types of
light tank used there; but the 45- and 47-mm. guns were far
more effective. It was clear by 1938 that heavier armour would
come along and would require a heavier gun to oppose it. This
I and others reported on returning from Spain, and I advanced
publicly and in print the development of a gun similar to the
German 88-mm. treble-purpose gun.

This weapon is not, as it has recently been represented to be,
an anti-aircraft gun converted in 1941 or 1942 to anti-tank
uses. Whether it was designed for all three purposes or not I
do not know; but it was being used by the summer of 1937 by
the German Condor Legion as anti-tank, anti-aircraft and field
gun. It is of very great value to any commander to have a
triple-purpose weapon of this type. As an anti-tank gun it is
heavy enough to knock out any normal armour; but anti-tank
gunners do not often have targets. They have to have their
weapons in place, and ready all the time, but enemy tanks
seldom come their way and then only for a few minutes.
Enemy aircraft pay rather more frequent visits, but move even
more rapidly out of range. Between these brief periods of
action the gun can serve as very effective field artillery.

It would have been natural for such a weapon to have been
developed in Britain, since any British operations of war are

likely to include landings from the sea, and it is much easier to land one triple-purpose weapon rather than three specialized weapons, each of which is useless or almost useless for the other purposes. Unluckily the armed forces of Britain lived so continuously within watertight compartments, during peace and the first part of this war, that even the three sections of the Royal Artillery responsible for anti-tank, anti-aircraft and field work scarcely seem aware of each others' existence. And if we possessed a triple-purpose weapon today – say our excellent 3·7-inch A.A. gun, better than the German 88-mm., mounted and munitioned for anti-tank and field work as well as for its primary purpose – we should probably need a year or two of discussion before a decision could be reached as to which section of the Royal Artillery should man and command these weapons.

How much the theory of warfare affects the development and use of weapons can be seen from the story of our own 25-pounder gun-howitzer. This weapon receives well-deserved praise. It is in fact the finest field gun in the world – for a war like that of 1914–18. It is a gun of unrivalled accuracy and endurance for the laying of barrages and for fire against unseen stationary targets. It has, on the other hand, less value than the 18-pounder it replaces against moving targets over open sights. The 18-pounder's ammunition could be of the cartridge type; in other words the ammunition would look like a much enlarged rifle cartridge, the whole of it in one piece, shell and propellant together. Ammunition for the 25-pounder is normally, as is well known, in two parts, charge and shell. This has certain advantages, but with other points of design slows down the rate of fire to such an extent that the 25-pounder is less effective against tanks than the 18-pounder, and much less than the German 88-mm. is, or our 3·7-inch could be.

While artillery is being dealt with, it may be in place to suggest that shrapnel is likely to regain some of its former importance. The shrapnel shell is fused to burst in the air not far above the target, and the burst releases a considerable number of heavy lead bullets which scatter down over an area wider than that normally covered by blast or fragments from a high-explosive shell. In 1914 most field artillery ammunition was shrapnel, and the proportion of H.E. was small. Then the armies went to ground, dug themselves into trenches. Shrapnel had not sufficient penetrative power to do much damage to men in trenches. It therefore decreased in importance and H.E. increased (until gas shells became, for field artillery, equally or more important). War today is fought far less in trenches than was the case in the past. During the decisive moment in many engagements the targets for artillery are either infantry in lorries or infantry scattering out from these lorries, or supply convoys, or other vulnerable vehicles such as tank transporters. I believe it may be found that shrapnel, because of the area over which it is lethal, is a better weapon against such targets than H.E. And in modern defence, when the main aim is often to keep enemy infantry from following up their tanks, I am certain that a number of field guns or triple-purpose guns firing shrapnel from forward islands of resistance is likely to delay the enemy infantry more than the same number of guns firing H.E.

Where, as in Russia, towns are strongly held, heavy artillery will come back into the picture. The Germans are already using a 16-inch howitzer and a 24-inch railway gun, against Soviet islands of resistance. But the main shape of war will remain too mobile for these monstrosities.

It is impossible to forecast with any certainty the effect of the use of poison gas on modern tactics. For the main defensive purpose stated, that of separating the attacking infantry from their tanks, persistent gas would seem to have some value, and to some extent it might slow up the process of modern attack and tend towards the negation of the blitzkrieg. On the other hand it might prove a valuable offensive weapon against islands of resistance so well fortified that projectiles have little effect upon them; it might have an even greater effect against guerrillas necessarily lightly equipped, and has been used by the Japanese extensively against Chinese armies and guerrillas who lack gas masks. The armoured vehicle can be made proof against gas, and so can the aeroplane; probably the effect of gas during the first period of its use in this war, if it is used at all, will be mainly to hamper or help in destroying those forces which are so conservative in equipment and tactics that they do not quickly make tanks, planes and troop-carrying vehicles gas-proof. The use of gas spray from the air may have an effect on civilian populations in areas where government propaganda has made gas into a terrifying neurosis; but its actual destructive value is small. Gas has not so far been used in this war on a large scale by our enemies because we possessed, until 1942, better facilities for making poisons and most of the rubber of the world. During 1942 we lost 90 per cent of the world's crude rubber resources, and our enemies are therefore in a better position than we are to 'armour' themselves against gas. It would be wise therefore for the question of the effect of gas on modern tactics to be more thoroughly studied. For example, as our equipment stands now it is not clear that fighter aeroplanes could be worked from aerodromes heavily sprayed with persistent gas.

The main processes in the development of modern infantry are three: motorization and close linking with armour; development of tank hunting and of defences against tanks; and approximation to guerrilla methods and tactics. Motorization has already gone so far that considerable forces can move at the speed of tanks; the weakness of these forces in action is that enemy machine-gun fire can make them dismount from their vehicles and can check their pace to that of men crawling. The Russians, trying to develop a close integration of tanks and infantry, have pulled armoured sledges of infantry behind their tanks on snowy ground, and have carried small parties of machine-gunners and automatic riflemen crouching on the actual tanks themselves. It seems likely there will be further development of both these ideas; various types of armoured infantry carriers will be tried out, and some form of bullet-proof shield behind the normal tank turret will make it possible for tanks to 'ferry' infantry forward at a rapid pace.

Infantry against tanks has not yet shown its full powers. This may be due to a theoretical misunderstanding of the nature of the job. Infantry has been given projectile weapons of various sorts, anti-tank rifles and light anti-tank guns, as its main weapons for use against tanks. I repeat that armour and in-fighting have always gone together; and I believe the theoretical conclusion from this is that infantry should be trained to tackle tanks mainly at extremely close ranges. They can only do so if trained to be practically invisible to tanks, which necessarily have restricted vision, at such ranges. And their main weapon against armour at such ranges is high explosive, either in the form of a heavy anti-tank grenade or the form of the anti-tank mine, or the form of some relatively heavy projectile which gets its effect on the tank by H.E. blast rather than by penetration of the armour due to high muzzle velocity. The solid bullet fired by the anti-tank rifle, or the solid shell fired by the light anti-tank gun, can only get its effect if it penetrates the armour and hits a vital spot; but H.E. blast can much more easily destroy tracks or running gear or blow a considerable hole in the armour. The bullet or solid shell must be propelled at a high speed; the H.E. projectile can be thrown by hand or projected relatively slowly. Against some types of tank flame is still effective; any tank must take in air, and where air goes in flame can go in.

Armour in the past has been conquered, as at Crécy, by projectile auxiliaries. It may be conquered again by such auxiliaries, plane or gun. But it may also be considered by something approximating to shock tactics – by the shock or blast of H.E., a new form of shock that could not be used in the past.

Men make the most complicated and costly weapons in order to cause an explosion near the enemy – great guns, bombing planes or torpedoes. And then they find against a dispersed or armoured enemy that only a small proportion of their ingenious and powerful projectiles secure any effect whatever. An infantry hidden either by camouflage and stillness or by smoke or by darkness can deliver explosive 'by hand' and are more likely to see that it gets to the right address. This new and – to some officers – rather frightening function of infantry is of course not in fact entirely new: the tank hunters of today and tomorrow are only the modern equivalents of the grenadiers of the seventeenth century. Like those grenadiers, they must be a picked and courageous infantry; but their job is possible and has already been done to some extent by men in Spain and Russia.

In a period dominated by armour and mobility it is impossible for infantry to protect themselves on the battlefield mainly by entrenchment. Their chief protection becomes invisibility, not only when enemy tanks are amongst them but when enemy planes are over them. That is the first reason why most of the methods and tactics of modern infantry will progressively be approximated to those of guerrillas. A guerrilla force is essentially invisible; it strikes from the dark, the jungle, the city's back alleys, or the difficult hills, and it retires or scatters to disappear immediately after it has struck its blow. All the conditions of modern tactics make it possible for a lively light infantry to fight in this way; swift and deep advances by mechanized and motorized forces, the swirl and ebb of attack and counter-attack through narrow channels between islands of resistance, have split up battle into a great number of isolated small engagements, in which the infantry unit that knows how to be as independent and as unpredictable as a guerrilla force has enormous opportunities. If infantry is considered, in such a battle, only as a defensive force against other infantry, or as moppers-up cleaning away the debris after an armoured attack has gone through, it cannot be fully integrated with the combat team of tanks and planes to which it is only an appendage or a target. But if on the other hand infantry plays its own and active part in the battle, in defence dangerous vermin hard to dig out of the seams of the soil, and in attack a stinging swarm that appears out of nowhere and vanishes as quickly as it appears – then infantry can once again become the most powerful factor in the combination that makes up an army.

Such an infantry will use smoke cover to a much greater extent than is normal at present. Infantry commanders of the conservative sort worry mainly about their own control over their men; smoke interferes with this control and units unaccustomed to it lose direction when moving through smoke. But continuous control over small units by commanders behind those units is no longer possible under any conditions of modern battle; and smoke or darkness are the only conditions which meet the theoretical necessity that I have stated – the necessity for infantry to rely for protection on invisibility rather than entrenchment. It was noticeable during the Japanese campaign in Malaya that the normal Japanese advance would cease by about three in the afternoon; the Japanese soldiers were then fed and rested until midnight or soon after, when their main infiltration movement would begin again. Using darkness to pass through our lines and encircle our units, they would be so placed at dawn that their fire could command not only our positions, from all angles, but also all ways of retreat from those positions. This is the typical technique of the guerrilla, and was doubtless learnt by the Japanese from the Chinese forces, which have carried guerrilla tactics and the approximation of infantry to guerrillas to as high a level of development as the Russians.

Infantry have a lot of street-fighting to do, in the years of the blitzkrieg and in the subsequent years that now – I suggest – are going to be shaped by a clash between the blitzkrieg and even newer ideas of war. The blitz attack aims at rapid irruption and then even more rapid movement, beyond the breaches created, to form the *kessel* or cauldron. Much of this movement occurs on roads; where roads cross there are towns and villages; if these are held by a determined infantry, the blitz is slowed down and the armoured formations are split up, channelled into smaller and smaller units, and cut off from lorry-borne troops and supplies following them.

Built-up areas in the last war used to be flattened by artillery, at great cost in time and shells. In a typical blitz attack they cannot be flattened; there is not enough time. They must be stormed. And they cannot be stormed by tanks. Tanks in streets are at a serious disadvantage. They cannot carry enough shells to destroy all the houses on each side of the streets. They cannot clean defending infantry from the rubble of ruined buildings, which gives this infantry as good or better cover than untouched walls and windows. A tank in a street is in a defile with enemy infantry at short grenade range on each side of it, and therefore it is constantly in danger.

During street fighting bombers are at a disadvantage; they cannot find their targets. So street-fighting is mainly infantry work; it is usually slow work; and the defenders have advantages over the attackers. When met by a defence that holds the built-up areas, and therefore the road junctions, a blitz force attacking must bring up its infantry and guns. The Russians held their towns; that is why the Panzer divisions, and the blitzkrieg itself, had to change and slow down in Russia. When such a defence is determined and skilful, as at Sebastopol, the whole town may have to be knocked down before it can be captured. This takes months; and the Germans had to waste precious months of spring and early summer on it in 1942. Where on the other hand defences round a town are linear – as at Tobruk in 1942 – the Germans break through these defences on a narrow front, drive straight to the centre of the built-up area, and work out from there to take perimeter defences from the rear.

Street-fighting and guerrilla fighting are linked in method and aim, as ways of resistance that can be effective against the blitzkrieg. Both have been taught, in Britain, by the school of military thought that began by teaching these methods of fighting to the Home Guard at the Osterley Park School in 1940.

The clearest proof that street-fighting can hold up the blitzkrieg is the siege of Stalingrad, which began after the text of this book was completed. At the moment of writing this note, inserted in the proofs, the siege continues; it is the biggest battle of its sort that has ever occurred. Tanks, planes, heavy guns and scores of infantry divisions have hammered the city. It is not a Verdun, for at Verdun the Germans never penetrated into the built-up area. The distinguishing characteristic of this fighting is that it is the defence of the built-up area, by street-fighting. The city is now being pulverized, as Sebastopol was; but it is built on softer soil than most of Sebastopol, and therefore the larger shells and bombs make craters from which the defence can be continued. There will be more battles of the shape of Stalingrad in the future, and fewer campaigns of the shape of Singapore – as soon as all our forces have learned street-fighting and the closely connected methods of guerrilla war.

Both Russians and Chinese have been able to maintain large and effective forces within areas nominally occupied by their enemies. The whole aim of modern techniques of war is to get men and weapons to effective points behind the enemy's main positions. It is with this aim that the blitzkrieg is designed, and very costly armoured machines made. It is with this aim that bombers and airborne armies are built; with this aim the parachutists come down. The guerrilla is 'there already'; he has his weapons at the point where the enemy is weakest. He strikes against enemy material where that material is stored or is in process of transport, and therefore cannot swiftly be brought into action, rather than against the same material deployed in the enemy's main positions and ready for action. The guerrilla also strikes at morale where morale is weakest, behind the picked units and the men securely armoured. A guerrilla force cannot dispose of the immense amounts of ammunition and other supplies that a more regular force in contact with its base can possess. Therefore it wastes far less ammunition; in its cloak of invisibility it gets to close quarters with the enemy and restores simplicity to battle.

The combat team that makes up the decisive force of the blitz type of attack is enormously powerful; but behind it must come the petrol lorries and the supply convoys, a great traffic jam of vehicles, mobile workshops, staff cars and all the rest. The aim of modern tactics is to use effective force against these soft parts of the enemy organism. A blitz force is like a man with head and trunk well armoured, but with no protection below the waist. It is unnecessary for any real soldier that I should complete this paragraph.

The pattern of warfare that can make the blitzkrieg obsolete, the emerging pattern of today, is in my belief that of the Peoples' War. This is a pattern of warfare which treats as principal the active linking of an armed population with an offensive striking force. Where possible this striking force will be patterned on and developed from that of the blitzkrieg; it will have tanks and planes and triple-purpose artillery and lorry-borne infantry and all the paraphernalia of mechanized war. But the weapons of the armed population will be cheap and simple to make: tommy guns, raw explosives for mines and grenades, machine rifles, etc. Civilian transport and civilian radio sets will help to link these forces to each other and to the striking force; they will use the cities as their hiding-places as well as the hills. They will fight in the streets as well as the fields. Guerrilla forces made from an armed population cannot be expected to be decisive without the backing of a striking force; the Spanish guerrillas could not defeat Napoleon's armies, though they could drain the strength of those armies. Only when Wellington's striking force was in play could the game be won. But the existence and work of the guerrillas make it possible for a relatively small striking force to defeat much more powerful armies.

The methods of the People's War have until 1942 only been used on the strategic defensive. By these methods the Chinese have resisted for five years the attack of an immensely more powerful enemy, supplied for almost all these five years by all the resources of the industrial democracies. These methods have helped to form in Britain a force, the Home Guard, that could be made capable of taking most of the duties of the defence of this country upon its shoulders, so that most of the regular troops in Britain could be used as a striking force on the Continent. In Russia the guerrillas have played not only a heroic part but a most effective part in the first year of 'the Russian glory'. But 'you ain't seen nothing yet': when the same methods are developed for the strategic offensive against fascism, the Peoples' War will seem, throughout Europe, more like an explosion than a campaign.

I suggested towards the beginning of this book that the underlying reasons for any revolution in the technique of warfare normally lie somewhat outside the developments inherent within warfare itself; such changes come when the peoples who make up a nation or several nations express themselves in a democratic or popular or revolutionary way. The democracy of the Greeks, of the 'barbarians' who broke the legion, Charlemagne's nation-state, the yeoman independence of the archers at Crécy, the released energies of American, French, Soviet and Nazi revolutions – all these, though difficult to equate or measure beside each other, have been the underlying reasons for revolutions in the technique of war. If my thesis is correct and the blitzkrieg is to be replaced by the Peoples' War, by a combination of guerrilla and mechanized striking force, this revolution in warfare will be caused by, and will take the shape of, a popular anti-fascist revolution throughout Europe and beyond Europe. To that the forces working and fighting against fascism in America or Russia, in the British Empire or China or ruined Europe, will each contribute in their own way. But no contribution can be made by those who hold to old ways of fighting, as they hold to old ways of living, because change seems alien and unpleasant and threatening to them. Change in technique, in tactics and strategy and supply and equipment, transport and training and every aspect of warfare – change is the only law that persists throughout warfare. If we are to survive and be victorious we must learn the ways of change.

**Part Two by J.N.Blashford-Snell**

## 11. The Third Armoured Period: Mobility

It is convenient to assume that Wintringham's third armoured period ends in 1945, with the coming of the Nuclear Age.

On studying the history of the First and Second World Wars, it is obvious that the latter was more mobile and on the ground was essentially a war of armour and artillery. This does not mean to say that the infantry was no longer important, but it was not the paramount striking arm, as it had been in 1914–18.

Armour and artillery possessed much greater fire-power and the mobility of the new weapons made more mobile tactics possible. This is illustrated at the outset of the war, by the blitzkrieg, which was in direct contrast to the defensive attitudes still fashionable at that time. Later the North African campaign showed how mechanized armoured forces could fight and manoeuvre over enormous distances, as if they were ships at sea. Finally, the liberation of Europe illustrated how quickly large armies could advance across a continent possessing a fine road network and bridges, built to carry heavy civilian loads in times of peace.

Scientific and technological progress made great strides and enabled the contestants to overcome many of the natural obstacles that had hindered earlier armies. In a simple work, such as this, it is not possible to consider all these advances and their effects, but there were certain general features which emerged and many have now become accepted, as a part of modern war.

The value of a balanced force, as seen in the successful blitzkrieg, was largely due to the excellent co-operation of the German dive bomber with the armour. The Stuka carried a relatively small bomb load, but it was delivered with accuracy by the aircraft in a vertical dive. A terrifying noise, emitted by a small device on the wings, added greatly to the initial psychological effect of this weapon system, especially as in the early days of the war anti-aircraft weapons were few and far between. The lessons of the Spanish Civil War do not appear to have been learnt by many soldiers outside the Third Reich.

On the ground, the improved performance of land vehicles increased the momentum of the battle and, at the same time, the fire-power and effectiveness of guns was such that the value of static fortifications was greatly reduced.

Throughout the war, both sides strove hard to gain more mobility for their guns, and artillery mounted on a tank chassis became known as the self-propelled gun. Unlike a tank, the gun could not traverse, but of course it fired a much heavier shell to a greater range than was the case with the contemporary tank armament. This meant that artillery could advance and deploy much more quickly than had been the case in previous wars. Difficult country, that had been churned into a morass in the First World War, was traversed with comparative ease by tracked vehicles. Although on the Eastern Front 'General Winter' made a formidable enemy.

Guns towed behind lorries were still used, for the prime mover or towing vehicle was also more powerful and thus able to negotiate difficult terrain, but the lighter weapons could be moved by air and were simpler to get across anti-tank obstacles.

The introduction of more reliable radio was perhaps one of the greatest boosts to the efficiency and effectiveness of artillery on the battlefield. Good radio communications enabled directions from observers, well forward and able to see the target, to be passed to the guns. Observers also used light aircraft to overlook the enemy positions. Like the balloons of earlier days, they were vulnerable but, being mobile, could usually get themselves out of a tight corner.

The range of artillery in the field was usually a maximum of 20,000 yards. Very long range guns were now replaced by aircraft and later by rockets. Even coastal artillery was declining at the end of the war and today is replaced by the guided missile.

Great emphasis was given in German tactical theory to the role of artillery, but in practice this arm was neglected and abused. Close gun support devolved by degrees on tanks, assault guns and mortars and of course, when they were available, dive bombers. There was a tendency to detach batteries of field artillery to isolated units, where the weight of their fire proved quite insignificant. The poor performance of the German artillery was a phenomenon of the war. Although its equipment and personnel were of good quality, it lacked the numbers and often the ammunition to be fully effective; but it was its poor tactical handling that was embarrassing when the Germans lost air superiority and their gunners had to counter the massed fire support of the allies. It may have been their desire to have close support for their armour that led to the situation. Indeed it was noticeable in 1942 that General Auchinleck began to counter German armour, by using field artillery dispersed in a similar way. On assuming command, General Montgomery reverted to concentrating his guns.

Commenting on his experience of the Russian Don Front in 1942, Marshal Chuikov says:*

Observing how the Germans carried out their artillery preparations against the 229th Infantry Division's sector, I saw weak points in their tactics. In strength and organization this artillery preparation was weak. Artillery and mortar attacks were not co-ordinated or in depth, but only against the main line of defence. I saw no broad manoeuvre with artillery cover in the dynamic of battle.

I was expecting close combined operations between the enemy's artillery and ground forces, a precise organization of the artillery barrage, a lightning-fast manoeuvre of shell and wheel. But this was not the case. I encountered the far from new method of slow wearing down, trench by trench. . . .

The German tanks did not go into action without infantry and air

*V. I. Chuikov, *The Beginning of the Road* (Macgibbon & Kee, 1963), pp. 33, 34.

support. On the battlefield there was no evidence of the 'prowess' of German tank crews, their courage and speed in action, about which foreign newspapers had written. The reverse was true, in fact – they operated sluggishly, extremely cautiously and indecisively.

The German infantry was strong in automatic fire, but I saw no rapid movement or resolute attack on the battlefield. When advancing, the German infantry did not spare their bullets, but frequently fired into thin air.

It is interesting to note that the efficient co-operation of the Luftwaffe displayed the familiarity of the pilots with the tactics of both sides.

In modern warfare victory is impossible without combined action by all types of forces and without good administration. The Germans had this kind of polished, co-ordinated action. In battle the different arms of their forces never hurried, did not push ahead alone, but fought with the whole mass of men and technical backing. A few minutes before a general attack, their aircraft would fly in, bomb and strafe the object under attack, pinning the defending troops to the ground, and then infantry and tanks with supporting artillery and mortar fire would cut into our military formations almost with impunity.

A degree of specialization, coupled with the conservative attitude of British forces, tended to form our thoughts in watertight compartments. The roles and weapons of anti-aircraft, anti-tank, field and coastal artillery were regarded as quite separate from each other and there were few, if any, all-purpose weapons in use by the allies. As is pointed out in the early part of this book, the Germans sought to produce a number of general-purpose guns and their most noticeable success was the dreaded 88-mm.

In 1939 Wintringham was advocating the production of a similar weapon or the conversion of the excellent British 3·7-inch anti-aircraft gun, but in spite of the grave shortage of effective anti-tank weapons, I can find no record of the 3·7 being used as such. It is easy to criticize the slowness with which the British army adopts new equipment and new methods; judgement after the event is easy. In the past the

British in general have not been war-like people and in times of peace our interests turn to fox hunting, football matches and fishing. There are more pleasant ways of employing one's hours of work and leisure than in the production of costly specialized weapons of war. Military theorists tend to be regarded as cranks or at least very odd. Inventors are beset with red tape and financial problems. The Treasury is the natural peace-time enemy of the soldier and they regard each other with the uttermost suspicion.

In fact of all these problems, it is to our credit that the tank was invented and first produced by the British. Military prophets saw a great future for the modern war chariot, but, alas, the British politician assumed that the long-awaited millennium had arrived and thus the British arms industry was starved of vital money and the establishments of our army were cut to the barest minimum. In spite of murmurings of war in the East, our government remained complacent in the belief that wars had ended.

The onslaught of the Nazi forces was heralded by the phalanx of fast-moving armour, that Wintringham has already discussed. Surprised and caught off balance, we were found wanting and it became necessary to rethink our tactical doctrine.

The British army had not long lost its horses and throughout the ranks of our cavalry there still existed an understandable spirit of dash, determination and courage. It was almost unnatural for such men to stand in defence or merely to attack weak, exposed enemy infantry. It seemed that their role was to fight off the enemy armour, which, by ill chance, was the strongest part of the German army. Headlong pursuits led our tanks onto the inevitable anti-tank screen, leaving unprotected infantry to the mercy of the German armour. The spirit of Prince Rupert at Edgehill and Marston Moor was still alive!

The guns mounted on our tanks as well as our anti-tank guns were generally too light, in the first part of the war, to stop the well-armoured Panzers. By the Battle of El Alamein we had new tanks with more powerful guns. Tactics had changed and no longer did we use the 'Thin Red Line' for defence, but constructed positions in great depth, through which Panzer phalanx could not penetrate. The same tactics were used in Russia, where even the technical superiority of the German tanks could not match the superior numbers of Soviet armour, the considerable depth of their defensive positions and 'General Winter'.

In the attack we had learnt to strike the enemy at his weakest point with our strongest weapons and to integrate the artillery, armour, infantry, engineers and air forces into a co-ordinated, all-arms assault force. Before leaving the question of tank tactics in the last war, it is worth dwelling on the German doctrine that was designed to be used on open desert or steppe.

If the enemy was entrenched in good positions, protected by natural or man-made tank obstacles, the German infantry would go ahead and clear the way. Their task was to destroy the enemy anti-tank guns and close-range weapons and give close protection to the sappers, who then lifted the mines and cleared the obstacles. Meanwhile the armour would give fire support from the rear. If the enemy defences were light and lacking in depth, tanks were concentrated to lead the attack at a decisive point and moment. The assault would be made by an all-arms team, backed by the necessary reserves. The first wave of tanks was given the primary task of penetrating the enemy line and destroying the artillery and infantry. Other arms assisted the tanks by eliminating anti-tank weapons. As the attack was made in depth, it was left to the tanks and anti-tank guns of the following waves to destroy the enemy armour and cover the movement of the leading echelon.

The attacks were normally launched on a narrow frontage (2,000–3,000 yards), with one unit behind another, in a manner similar to that used by Napoleon. Alas, we lacked the anti-tank weapons to follow Wellington's counter tactics.

The Germans sensibly declined to attack anti-tank screens head-on, and believed in engaging enemy armour only when they had superiority of armament and range. Movement was covered by fire and combined arms training was considered of the uttermost importance. The infantry moved in lorries or armoured personnel carriers and learnt the art of infantry and tank co-operation.

The Germans were quick to develop tactics for eliminating the British armour in Libya. Unsupported tanks were drawn onto a screen of anti-tank guns and destroyed in detail. Some guns were sited to fire into the more thinly armoured flanks of the attackers, and on occasions they were allowed to pass through the outer screen of anti-tank guns, before being destroyed from behind. The Persian defeat at Arbela had been forgotten. Throughout the war, and even in 1970, there is the greatest competition between the tank and the anti-tank weapon.

In the later phases of the war, special circumstances caused the development of various special tanks. These became known as the 'Funnies' and were designed to overcome obstacles. Bridge-laying and the filling of anti-tank ditches to enable our armour to cross were most important. In Europe and Burma, conditions were very different from the Western Desert. Progress in Italy and Holland would have been almost impossible without some device to enable the momentum of our armoured advance to be maintained. Other Funnies included the A.V.R.E. (Armoured Vehicle Royal Engineers), which could bridge obstacles under fire and also carried 40-lb. H.E. charges that could be fired against pill-boxes, from a spigot gun fitted on the turret. These mutually supporting pill-boxes often formed the basis of a German defensive system and they were extremely difficult to destroy with a normal gun. In the Pacific similar emplacements built of earth and bamboo logs caused heavy casualties to the American forces and were only overcome by courageous and costly attacks, using high explosive and flame throwers.

Another Funny was the crocodile, a Churchill tank fitted with a flame-thrower. It was first used in north-west Europe shortly after D-Day and the Germans greatly feared this

dreadful weapon. Its value as a morale destroyer was considerable and the sight of this twentieth-century dragon was indeed
terrible to behold. The few men actually caught in the rolling
tongue of liquid fire were instantly incinerated, and partly for
certain humane reasons flame-throwers went out of service
shortly after the war. Today they are replaced by the equally
terrible, but less discriminating napalm bomb.

Amphibious tanks, 'flails' and 'soft-sand track layers' were
used on the Normandy landings to get armour ashore quickly
and clear a way forward, through what was believed to be a
great obstacle belt. The flail, or crab as it was sometimes
known, cleared paths through minefields by chain lashing out
from a power-driven spindle, carried on a boom at the front of
the tank. As the mines detonated they would damage the
chains but the tank remained intact. Thus a path was cleared
through an anti-tank minefield, which permitted the armour to
advance with the infantry. On the beaches, areas of soft sand
were paved over by a flexible track, which unrolled from a
large drum fitted on the Funny.

In reality, the beach defences were not as formidable as had
been feared, but the Germans adopted a typical defence in
depth, which might have been more successful had it not been
for the sheer weight of bombing, shelling and airborne assault,
coupled with the activities of the French resistance, which
accompanied the landings.

Nevertheless, the Funnies were greatly appreciated and it is
worth noting that the Americans, who had been offered these
strange devices, had rejected them. Brigadier Peter Young in
his book *World War 1939–45; A Short History* (Barker, 1966)
says:

The American planners had been offered large numbers of the
Funnies which their British allies had developed to deal with pill-
boxes, barbed wire and mines. These were Shermans converted for
specialized purposes. Some were equipped to flail a way through
minefields, others to lay tracks or to bridge anti-tank ditches. Some
had flame throwers or huge charges for destroying pill-boxes. Such
devices might be all right for the cautious and war-weary British:

the Americans preferred to see what could be done by straight-forward frontal assault. Dieppe might never have been fought!

The Bocage country to the south of the beaches was ideal for defence. Thick, almost impenetrable hedge-rows channelled the movement of vehicles to the narrow country roads and lanes. Every farm and every hamlet was contested and heavy tiger tanks lurked in the coverts and orchards ready to deliver a high-velocity, armour-piercing shell at point-blank range. Anti-tank guns of every sort and size, mines and bazookas made the job of the attacking armour extremely unpleasant. Gone were the days when our tanks had roamed, almost like battle fleets, in the Western deserts. Now no tank dare move without the support of infantry and it was fortunate that the allies had established considerable air superiority. Even so, there were tragic errors of recognition and, as is now so common an event in Vietnam, our aircraft sometimes attacked its own troops. A fighter flying at 400 m.p.h. finds it difficult to distinguish friend from foe, especially in the lush countryside of Normandy that fateful summer, where opponents lay 50 yards apart.

Once the allies had established their foothold in Europe, thinking Germans realized that the war was lost. Indeed, some had realized that the tide had turned in the Solomons, at Stalingrad and El Alamein in September/November 1942, when Churchill had said, 'This is not the beginning of the end, but the end of the beginning'!

The German soldiers, fighting without air support, quickly found themselves short of all the essentials of modern war. The marauding allied aircraft bombed, machine-gunned and rocketed all attempts to reinforce or resupply the teetering forces fighting to contain the swelling bulge in Normandy. The allies, with short lines of communication and the Mulberry artificial harbours, received abundant supplies and a constant flow of fresh troops, to prepare them for the inevitable breakout. It was merely an example of concentrating one's forces in preparation for a strike at a decisive point. Montgomery, with his experience of El Alamein, knew the tactics well. The 1914–18 War had become a stalemate in Europe, because contemporary transport failed to keep the armies sufficiently supplied beyond the rail-heads. It was not only the machine-gun and the wire that prevented a decisive victory, but the immobility and the inability to follow up and exploit success quickly. In the Second World War, similar problems led to the German Defeat at Stalingrad.

In 1944 motor vehicles were more reliable and more powerful. Horses were not used by the allies, but they were found in many units of the German Army, especially on the Eastern front. At Falaise the roads were jammed with burning vehicles and the decaying carcasses of horses, trapped inescapably in the now famous pocket. The ubiquitous jeep and other four-wheel-drive vehicles, fitted with lasting tyres, helped the allies to advance quickly.

Tanks now had far greater mobility, protection and fire-power and thus the other arms had also to have mobility and some protection, in order to exploit this advantage and keep pace. Thus there arose a breed of vehicle known as the Armoured Personnel Carrier (A.P.C.). The idea was not new and indeed a type of A.P.C. has appeared time and time again throughout history. It should not be compared with Hannibal's elephant, but more with a self-propelled form of the Roman soldier's shield. The purpose of the A.P.C. was to protect the soldier from missles of the lighter variety (e.g. bullets and shell splinters) and enable him to close unscathed with the enemy. Thereafter he dismounted and fought in the normal way. Some German A.P.C.s had provision for infantry weapons to be fired from them, either through weapon slits or from the top of the vehicle. Without special mountings, it is not easy to fire small arms accurately from a vehicle moving rapidly over rough country. However, the shock effect is tremendous, and if one can imagine being attacked by hordes of these mobile forts, supported by apparently invincible tanks, the prospect is not pleasant. If morale has already been lowered by air attack, shelling, privation and a shortage of ammunition, only the stoutest of hearts will stand and fight. Another step has been taken in the consolidation of the third armoured period.

In order to defeat the armoured attack, defence had once again to be based on the skilful use of ground and obstacles. If the obstacles were to be effective, they needed to be covered by observed fire and these tactics required that weapon accuracy and range should be increased. The obsolescence of permanent fortifications in the face of the blitzkrieg technique had caused emphasis to be placed on the principles of all-round defence, mutual support, concealment from ground and air view and protection from artillery bombardment. Bitter lessons were learnt and a dear price was paid for the failure to cover the British minefields, in the Western desert, with fire. The ideal defence in 1944 might have been a web system of strong points and obstacles sited in great depth. Mobile counter-attack forces should then be available to operate over ground known to them and strike at the penetrating attacker. The German-defended river lines in Italy proved extremely effective in blocking the allied advance and, had sufficient landing craft been available, obviously it would have been better to have outflanked them on the sea.

Hitler had forced his armies to delay too long in Normandy, instead of withdrawing and establishing a new defensive position on the Seine. Indeed, after their disastrous losses in France, it is amazing that the Germans succeeded in fighting on two fronts and regrouping sufficient forces to enable them to launch the Ardennes offensive that December.

However, although the blitzkrieg technique achieved success in Europe it was of much less value in battles like Leningrad and Stalingrad. Here stubborn resistance by the Russian Army, the severe climate and difficulty of movement all contributed to causing Hitler the greatest disaster of his brief command.

The contribution of sea power must not be forgotten in this work, which is largely a discussion of land tactics. Without the support of their navies, the allies would have had no hope of carrying out successful amphibious assaults, as was done in the Pacific and Europe. In those crucial moments, whilst the army is establishing itself ashore, the protection provided by carrier-based aircraft and naval gunfire is vital. Even so, the effect of heavy and prolonged bombardment on defenders living in deep, well-constructed and tactically sited bunkers is not as great as might be expected. The terrible casualties suffered by unprotected infantry, as they landed on the numerous Japanese occupied islands, bear witness to this fact. In an attempt to minimize these casualties, special vehicles such as the Amphtrac were developed. This was a lightly armoured amphibious vehicle that could carry its occupants across the exposed beach and, it was hoped, into a relatively safe debussing area beyond. Perhaps the hovercraft may be able to do this in future.

As the need for amphibious assault became apparent, there was a requirement for detailed reconnaissance of the beach defences. Low-flying aircraft were of great value for photographic, aerial reconnaissance, but the real need was to get a man onto the beach. Thus men were equipped with underwater breathing apparatus, to enable them to swim in undetected from the sea. Man had taken personal warfare beneath the sea. The Soviet interest in undersea warfare may be an ominous portent.

## 12. The New Mobility

Although this book is primarily concerned with the land battle, we cannot consider any modern campaign without studying the relationship of the land, air and sea forces. Indeed the close relationship between these forces was well illustrated throughout the war. It had been thought that the Second World War would be, above all else, an air war, but this did not prove altogether the case.

Throughout history, nations have frequently prepared for a kind of war in peace-time that in fact has failed to occur in practice. Thus too much emphasis may have been placed upon the importance of bombing.

Before the war, theories put forward by the Italian Brigadier Douhet strongly recommended strategic bombing, but he saw no great future for aircraft as tactical weapons. Douhet believed that air superiority must be obtained, but he thought that this would result from bombing as opposed to aerial combat. He saw the land and sea forces initially on the defensive, whilst the air force massed and prepared for the battle of the skies. He believed, with some foresight, that the attacking side could win enormous advantages by unleashing its air forces in a massive surprise onslaught at the outbreak of war. However, Douhet was preoccupied with his consideration of the bomber and failed to credit the fighter as an effective method of defence. He overestimated the effect of offensive bombing and many of his theories were found wanting at the Battle of Britain.

Although air superiority was shown to be necessary for success, the air arm alone could not win decisive battles, although an exception is perhaps the Battle of Britain.

The strategic bomber was simply a means of delivering a powerful weapon, with reasonable accuracy at a great range. Unlike the gun or early rockets, the bomber could seek out its target, but it suffered from the disadvantage that it too was vulnerable.

Although defeated at the Battle of Britain, the Luftwaffe was still an extremely powerful air force and the effect of its bombing was felt throughout Britain and many other countries. Most of our major cities suffered considerably and yet the Germans failed to achieve their strategic aim and though industry, communications and shipping were frequently and heavily bombed, the British people were not subdued.

It must be remembered that the Luftwaffe was designed as an army-cooperation tactical air force, which was perhaps why it did not do well in the strategic bombing of Britain. In contrast the R.A.F., except for Fighter Command and its single limited defensive purpose, was designed in obedience to Trenchard's belief that air power was a separate entity and could decide a war on its own, and hence until about 1942 gave the army and navy little useful aid.

After Dunkirk, Britain had only one effective means of striking back at Hitler. Her army was defeated and her naval resources stretched, but she possessed a small, if somewhat obsolescent bomber force. This was the nucleus of the now famous Bomber Command. Initially it had to be preserved and therefore it was good tactics to make concentrated attacks on one target at a time. Whilst following these tactics, our bomber force expanded until after the historic 1,000-bomber raid on Cologne, the Germans were forced to redeploy many of their fighters to meet the growing threat. To do this, they had to take Luftwaffe forces away from their role of co-operation with the army and deploy them to defend vulnerable cities and industrial areas.

As the war progressed, bombers became able to carry greater loads and fly to longer ranges. Their greater endurance enabled them to play an important part in guarding our shipping and seeking out the ubiquitous U-boat. Although they could devastate cities and destroy industry, science had not yet produced bombs that could penetrate the enormous thickness of concrete, which shielded the U-boat pens and the bunkers of the Atlantic Wall. Nevertheless, slowly the German nation began to crumble beneath the relentless offensive. This was not achieved without many losses to the allies. The slow-flying bomber was vulnerable to anti-aircraft fire and the German fighter. On 14 October 1943, 291 Flying Fortresses attacked a German ball-bearing factory. For the last stage of their flight, they were beyond the range of their protective fighter escort and although the target was heavily damaged, the Americans lost sixty aircraft to German fighters. Well over half the survivors that returned to Britain had been damaged. The result of this disastrous raid was the immediate production of a long-range, high-speed fighter known as the Mustang. This new weapon reversed the situation and permitted long-range, daylight bombing to continue.

The use of heavy bombers in a tactical role was less successful. In Italy and in Normandy, great quantities of bombs were delivered on forward German positions and it was amazing to those who beheld the holocaust that the enemy was still capable of stout resistance once the bombing had stopped.

In fairness to the airmen it should be pointed out that perhaps part of the failure of such combined operations (e.g. Operation Goodwood in Normandy) may have been the relatively slow follow-up by ground troops. Although it seemed rapid at the time for the army to be attacking the enemy within an hour of the bombers' departure, an even more speedy response, such as is possible with parachute troops, may well have been the answer. This presupposes that such airborne forces and aircraft are available at the time.

At Cassino the limitations of tactical bombing were illustrated once more. As the allies advanced from the south of Italy, following the Italian capitulation, they were brought to an abrupt halt in the shadow of the monastery at St Benedict. This great building stood on a prominent hill, blocking the road to Rome. Here the German troops had constructed a network of immensely strong bunkers and fortified positions stretching across Italy. This barrier to the allied advance was known as the Gustav line. To break through the defences, it was decided that the monastery must be removed, and on 15 February 1944, 254 bombers delivered 576 tons of bombs onto the huge building. Although the target was destroyed, the debris merely collapsed on top of the German bunkers and, if anything, made them even stronger. More bombing attacks and enormous artillery bombardments followed until eventually, on 17 May, after heavy fighting, the Polish Corps captured what remained of Monastery Hill.

The bomber, although it contributed very largely to the strategy and in some cases to the tactics of this war, was no more the ultimate weapon than was the U-boat. Bombers alone could not destroy well-prepared defensive positions and only after a very long time could they wear down the industry and spirit of a nation. Their limitations may be summarized as their range, their vulnerability, the accuracy of their bombing and the power of their bombs.

Nevertheless it was carpet bombing by heavy bombers that had disappointing results at Cassino and in Normandy. Tactical bombing by light and medium bombers was much more effective.

The value of an air force was also its ability to deliver men, equipment and supplies to otherwise inaccessible areas and often gain surprise.

A foretaste of things to come was seen in May 1940 when the Germans were pouring into the Low Countries. A vital part of the allies' defence plan was the apparently impregnable Belgian fortress of Eben Emael. A well-trained combat team of parachutists, including assault engineers, descended on the fortress and within thirty-six hours had forced the garrison, still 1,100 strong, to surrender. The garrison had only lost 100 casualties and had the advantage of considerable pre-planned supporting fire from neighbouring forts. Nevertheless the attackers' briefing and preparation, much of which had been done on a full-scale model, was so thorough and their arrival such a surprise that they overcame the strong point with few casualties to themselves.

In 1941 the Germans carried out the largest and most ambitious airborne operation to date. The island of Crete was held by Major General Freyberg, v.c., the famous Commander of the New Zealand Division. Freyberg expected an airborne attack and had done his uttermost to prepare to meet such an assault. However he had insufficient ground forces and no fighter support when on 20 May 1941 the Germans invaded.

The initial objective of the Luftwaffe was to destroy Freyberg's anti-aircraft guns. This was quickly achieved and followed by well-trained parachute troops, supported by waves of gliders carrying more ground troops, and heavy weapons. The defenders fought heroically and inflicted terrible casualties on the Germans. An attempt to support the operation by sea was stopped by the timely appearance of the Royal Navy, and even though the British paid a heavy price the Germans made no further attempts at seaborne invasion. On Crete the airborne forces gradually succeeded in gaining a foothold and as soon as they captured an airfield they were reinforced by transport aircraft at the rate of twenty an hour.

Eventually the British were forced to evacuate the island, but the victory had cost the Germans dearly. They had lost between 12,000 and 17,000 men and 170 of their troop-carrying aircraft. Some units had lost all their officers and in others only a handful of men had survived. It had proved so expensive that the Germans never repeated such an operation.

On the other hand isolated garrisons supported and supplied by air could hold out for long periods against superior odds. A successful example of air maintenance of a beleaguered garrison was during the German Ardennes offensive of December 1944, when the 101st American Airborne division held Bastogne like a rock in the path of the 5th Panzer Army. During several days of bitter fighting, it was air supply that saved Bastogne. In Burma it was even more essential. Here was a country largely covered in dense, impenetrable jungle and intersected by great rivers. Mountains, swamp and paddy field added to the difficult nature of the terrain. The roads were few and in poor condition, becoming quickly mud-bound under the torrential monsoon rain. In the forward areas the jungle canopy permitted infiltrating forces to approach the vital supply lines and attack the convoys. Thus air supply on an enormous scale was used by the allies.

With air superiority to keep the Japanese fighters at bay, the U.S. Air Force and the R.A.F. lifted 615,000 tons of supplies, flew in 315,000 reinforcements and took out 110,000 casualties. Our offensive in the Arakan depended in the end on defensive boxes, which held out simply because they were supplied from the air.

Air supply is a costly business and can be severely limited by bad weather. If the aircraft is able to land, more stores may be delivered than by parachute dropping. Air landing of supplies and reinforcements had the added advantage that casualties and, if necessary, P.O.W.s may be evacuated from the battle area. However, it would be mistaken to assume that with air supply a garrison may hold out indefinitely, and the fate of the French at Dien-Ben-Phu is evidence of this.

The delivery of men by air was the subject of considerable study by the Soviet Army before the Second World War and numerous methods were tried with varying degrees of success. One hair-raising type of delivery involved a small number of men being placed in a closed metal box on wheels. This was released from the aircraft, flying relatively slow at ground level, so that the container acted as a false undercarriage and eventually slowed to a halt. As far as is known, the method was not used in war, but it is interesting that today similar ideas for low-level delivery have been adopted. However so far only stores are landed thus!

If no airfield was available at the destination, men had to be delivered by parachute or glider. The use of both methods provided new means of outflanking or enveloping one's adversary and the use of the parachute to deliver large forces and small units behind the enemy front is now well understood. Similarly troops could be landed by glider and this had the advantage that less-specialized training was required and provided suitable ground was available, relatively large formations could be delivered, together with bulky and awkward equipment. The parachute force relied entirely upon resupply by parachute until, as was the case in Crete, it could capture a suitable landing ground.

Airborne forces generally, can only hold out for a limited period, as they lack heavy supporting weapons and armour, both of which are essential in modern war. Their supporting aircraft are highly vulnerable to anti-aircraft fire and fighter attack, especially in the final stages of the run-in to the dropping or landing zone. The epic battle of Arnhem has gone down in history, but in spite of the valour and great sacrifice by the parachutists there was no hope whilst the Germans held off the attacking columns of ground forces, with which it was planned to link up. As the battle ebbed and flowed the R.A.F. tried desperately to bring in supplies; tragically in the confusion many of these fell into German hands. Unfavourable weather, radio communication difficulties and the swift reactions of the enemy all contributed to the disaster and highlighted for posterity the limitations of airborne forces.

In practice, although gliders were widely used in Europe and Burma, they were not a great success. The high percentage of crashes on landing caused heavy casualties to their occupants, and indeed it was largely due to the courage and skill of the glider pilots that the operations were at all successful.

One of the most daring and successful glider-borne operations of the war was the capture of the bridges over the River Orne and the canal at Benouville in the early hours of D-Day 1944. Six platoons of the Oxfordshire and Buckinghamshire Light Infantry under Major R. J. Howard flew in from Britain and in spite of the darkness of the night landed almost on top of the bridges. In fact one glider came to a halt with its nose through the barbed wire of the German strongpoint at the end of the canal bridge. The seizure of these bridges was of the uttermost importance to the allied landings that took place later that morning.

At the same time parachute assaults were being made throughout the region, but owing to faulty navigation and a great deal of sheer bad luck these airborne forces were scattered over very wide areas. Many parachutists were lost in swamps, others were captured and overall the casualties were very heavy. It is to their lasting credit that the depleted units achieved all they did.

Today the helicopter has replaced the glider and to some extent the parachutist. It is most interesting that this apparently cumbersome, slow-flying means of transport is so successful, even in the face of a determined enemy. The reason may be that the glider lacked any form of motive power once released from its towing aircraft, whereas the helicopter is powered and can land vertically in a small space and take off again. The introduction of the sky cavalry has undoubtedly given much increased tactical flexibility to the armies of the seventies. This will be discussed more fully in a later chapter, but it must be obvious that for a soldier to be independent of the ground, to rise and fall at will and fire his weapons from this aerial platform has considerable advantages. Perhaps we detect the end of an armoured period!

As aircraft became more effective, so the ground forces needed weapons with which to destroy them. Naturally the fighter operating in the same environment was the primary enemy of the bomber, but ground troops and important installations required protection. The simplest anti-aircraft guns quickly developed into more complicated and expensive weapons. Radar was a vital accessory that could detect the approaching target and predictors calculated the course, height, speed and future position of the aircraft. In theory, if it was in range, you could not miss, but in practice many thousands of shells were fired for every one that scored a hit. Sometimes shell splinters falling back to earth were a serious hazard, especially in densely populated areas, such as London. Attempts were made to meet mass bombing raids by using batteries of rockets. However at this stage they were not guided by remote control or any form of homing device and their accuracy was questionable. As the war progressed, the bombers became more powerful and were able to fly faster and higher. The war ended with the bomber well ahead of the anti-aircraft gun.

At low level there was parity and here the fighter bomber was engaged by the rapid-firing 'pom-pom' and Bofors gun. But even here the speed of aircraft was having its effect and it was becoming very difficult for gun crews to traverse their weapons sufficiently quickly even assuming they had warning of the aircraft's approach. This was overcome by the introduction of electrically powered elevating and traversing motors and also by the fitting of more refined sights and improved early-warning radar. A new era was dawning and new weapons would be necessary if the speed of aircraft were to continue to rise. Thus was born the missile that could be guided to the bomber by radar or its own homing device.

## 13. The Birth of the Missile

As the strategic bomber was a means of delivering a weapon, it is perhaps not surprising that men of ingenuity sought to replace it with an unmanned aircraft, capable of a similar amount of destruction. The quest for new weapons goes on in peace and war. During the Second World War there was an approximate balance as to the initiative, held in some fields by the allies and in others by the Germans. As new discoveries are made, so leaders of nations and generals hope to produce a weapon which will be decisive in terminating the conflict in their favour. The crossbow, gunpowder, the machine-gun, gas and the tank have all, in their turn been considered as the knock-out weapon of their time.

Hitler pinned much faith on his flying bomb. The first attack was launched at dawn on 13 June 1944. Four small, pilotless aircraft propelled by a simple engine, employing the jet principle, exploded on British soil. They were crudely made of reasonably cheap materials and were launched from ramps in occupied Europe. Their speed was slow by today's standards, as it is believed that their cruising speed was approximately 350 m.p.h. Therefore the tactics employed against these weapons were to shoot them down with anti-aircraft guns or fighters and the fact that they could take no evasive actions was a serious disadvantage. Indeed, it is known that, at times, fighters were able to overtake the flying bomb, or V1 as it was sometimes known, and by positioning their wing tip under that of the pilotless aircraft, turn it off course so that it exploded far from its intended target. These weapons were completely without guidance once they had left their launching ramps and their accuracy was poor. However with a warhead of approximately 1,000 kg. of high explosive their destructive power was immense. They undoubtedly had a detrimental effect on the morale of the population, and aircraft were diverted to seek out and destroy the launch sites which, because of the limited range of the bomb, were mostly within striking distance of the Channel coast.

The V1 was the shape of things to come, and on 8 September 1944 the first V2 fell on London with a shattering detonation. It came at high speed from great altitude, plunging almost vertically downwards at a speed far exceeding that of any aircraft. There was no question of any sort of interception, other than by some static obstacle such as a barrage balloon. Its warhead was of similar size to its predecessor's, but it was a much greater threat than any aerial weapon known at that time. The production and operation of this complicated, unguided rocket must have played havoc with the German war effort and its lack of pin-point accuracy cast some doubt on its value. Nevertheless, it was fortunate that German research into guidance systems had not reached the degree of efficiency to warrant their inclusion in the missile. When we consider the destructive power of the inter-continental missiles of today, it may be of interest that over 1,000 V2s fell on London and south-east England, killing or injuring over 9,000 civilians, in the six months before the launching sites were over-run by the advancing allies. These were simple weapons with a maximum range of 200 miles and although the German scientists working on this irresistible weapon expected them to be armed with bacteriological warheads, in fact they carried only conventional explosive. In the last twenty-five years, science has advanced to a stage where, by using the descendants of the V2, we are able to land men on the moon or cruise past Mars. This does not bode well for the chances of a civilian population in a future conflict, especially when there are almost no air-raid shelters in existence.

## 14. Gideon Rides Again

In the Second World War many unorthodox units were raised
for special purposes. The missions carried out by these fearless
bands of determined men have become legendary. Although
few of the original units remain in being, the respected names
such as the Special Air Service (S.A.S.), the Commandos, Long
Range Desert Group (L.R.D.G.), Wingate's Chindits and Pop-
ski's Private Army (P.P.A.) are well remembered. The special
forces were of two types, those intended for offensive raiding
and those whose task was primarily to gather intelligence.
However, there was obviously a certain degree of overlapping
of reconnaissance and raiding. Most of the units operated as
small parties, although the Chindits, the Commandos and, later,
the S.A.S. became major units of considerable strength. The
Chindits fought entirely in the Far East under their energetic
controversial commander and founder Orde Wingate.

As it is said that Major General Wingate was considerably
influenced by Tom Wintringham, it is perhaps fitting to use
him to illustrate the individuality of the leaders of such groups.
He was an unusual man, who commanded the confidence of
other unusual men. Wavell, the soldier/scholar, believed in
him and had allowed him to organize a Jewish volunteer
militia to fight the growing Arab terrorism in Palestine. Later
he proved his genius for organizing unconventional warfare
and guerrilla forces in Ethiopia where, in a brief campaign, he
did much to bring about the downfall of the Italian Army of
occupation. He is still remembered in that mountain fastness
by the proud Amharas, who believe that it was Wingate and his
staff of British officers who were largely responsible for the safe
return of their Emperor. Churchill found himself in harmony

with Wingate's daring and powerful mind and Mountbatten also encouraged him.

At times he suffered fits of the deepest depression and at one time attempted suicide. Wingate would not suffer opposition. Every obstacle was a challenge and every setback spurred him to greater efforts. It may have been this stubborn determination that led to his death, for he was killed in Burma, whilst flying through terrible weather in a mountainous region.

General Slim, his superior officer and commander of the 14th Army did not always see eye to eye with this rebellious character. However, Slim wrote after Wingate's death,

Wingate had clear vision. He could also impart his belief to others. Above all, he could adapt to his own purpose the ideas, practices and techniques of others, once he was satisfied of their soundness. To see Wingate urging action on some hesitant commander was to realize how a medieval baron felt when Peter the Hermit got after him to go crusading. Lots of barons found Peter the Hermit an uncomfortable fellow, but they went crusading all the same.

The tactical conception of the Chindits was of a number of columns, each approximately 300 all ranks, supplied by air, self-supporting for periods up to a week and therefore independent of any line of communication (L. of C.) or supply. Their role was to harass and cut the Japanese L. of C. There were two separate expeditions. The first, in early 1943, was of Brigade strength, but the second in March–May 1944, was carried out by a force in excess of a division, of which half were flown in by Dakota and glider to landing strips in the middle of Burma.

Wingate's sudden death came during the second expedition and left behind him a growing number of critics who could not see the real value of the Chindit operations. The sight of bearded, untidy men was contrary to the British way of life, even at this stage of the Second World War. The thought of the abandoning of sick and wounded to the enemy was even more contrary to the traditions of the British way of making war. However, much useful experience had been gained on air supply and the whole concept of achieving tactical surprise with air mobile units was demonstrated. In the second expedition, the Japanese had to withdraw large forces that were desperately needed at the front, to deal with the threat to their L. of C. Although the Chindits suffered terrible casualties, they showed that the European was not inferior to the Japanese and by so doing contributed to morale in the army and at home.

Indeed, the now famous raids carried out by the Commandos had a similar effect. These amphibious soldiers were trained to operate in small bands or as a larger unit. They were highly skilled fighters, using landing craft or small boats to reach the enemy, as Wingate used aircraft. Their weapons and equipment were light and portable. Their leaders were men of dash, determination and courage. Being somewhat unconventional they easily adopted flexible tactics and new ideas, designed to use their mobility to the best advantage. Against an enemy using static methods of defence, they achieved considerable success. It is interesting to note that rarely did the Germans carry out raids of a similar type. However, they did have one notable success against the allies, when they raided St Malo from Jersey, in the closing stage of the war.

In general the Germans showed little interest in such tactics, whereas the British had a taste and flair for 'Bohemian' styles of making war in small groups. Possibly because they felt that they simply had to hit back with something! Yet in spite of this they failed to infuse their more conventional large formations with the enterprise and dash shown by their 'private armies'.

By the end of the Second World War, the S.A.S. had increased to five battalions of parachute trained troops, controlled by a Brigade Headquarters. Their object was to operate behind the enemy front line in small, controlled parties, in uniform, in an offensive role. Today commandos and S.A.S. type units exist in many countries. Since 1945 the Royal Marine Commandos have proved their value in many campaigns all over the world. Britain still has a regular S.A.S. unit, and similar U.S. Green Beret special forces operated in Vietnam.

Since 1945 the politico-military war waged by the Communist Bloc, together with the peripheral type of war fought by Britain, has increased the demand for units capable of operating for extended periods in small parties.

As the cold war continues, the demand for special forces will probably grow. Today they may be summarized as units where men are trained to operate for long periods in small parties, in areas controlled or occupied by the enemy. They have an ability to arrive by a great variety of methods, including parachute, small boats, or over hitherto impassable country. Their radio communications are first-class and they are well trained in the individual skills such as signals, medical duties, demolition and gathering intelligence. Many of these soldiers speak a foreign language, and above all they are extremely fit and able to cope with wide extremes of climate and terrain. They can provide 'stay behind' parties in nuclear war.

However, it should be understood that they have limitations. Men of such calibre are few and far between and they should not be exposed to taking heavy casualties, such as might occur if special forces, with their limited fire-power, were asked to hold ground against an organized and determined enemy.

Mobility, flexibility and speed of deployment are their great advantages.

# 15. Ho Chi Minh and All That

The activities and example of the special forces in the Second World War gave a great boost to many guerrilla movements. The Western allies went to some length to supply advisers, arms, radio sets and funds to groups of organized resistance fighters, who would turn against the Axis powers. To a lesser extent, the occupied peoples of Asia were similarly encouraged to wage war upon the Japanese. Thus in 1945, there were many 'civilians' who had been trained in espionage, sabotage, clandestine operations and to gather basic intelligence. An esprit de corps sprung up amongst those who had risked all in these irregular armies. What was more, many of them did not hand in the weapons and equipment supplied for killing Germans and Japanese. It was perhaps typical of naïve Western thought that they should have been expected to return readily to their former more peaceful occupations.

In Europe the enemy was utterly defeated and naturally men longed to return to their pre-war state of reasonably comfortable living. A reunion, a party with the boys or perhaps a visit to some especially hallowed 'battlefield' was all that most European guerrilla fighters needed to remind them of the war, but in Asia the situation was very different. The people had little to look forward to under the new regimes and, as the Japanese dream of imperial expansion writhed in its death throes, new oppressors arrived on the scene.

In Indo-China the struggle quickly turned against the French. Throughout Asia the white man was no longer regarded as invincible and, although he had triumphed in the end, the indigenous people knew that the Europeans could be defeated in battle by an Asiatic nation. They had witnessed the humiliation of their masters and in some areas their respect had grown correspondingly less.

Guerrilla war is not new to the world – its name is taken from the Spanish 'Guerrilleros' of the Peninsular War. Encouraged and supplied by Wellington, these Spanish irregulars put up a fanatical resistance to Napoleon's forces, killing an average of a hundred French soldiers daily. They were ragged, dirty and ill-armed, but they fought with a ferocity rarely seen on the more conventional battlefield. They fought a war without knowing the rules or accepted practices of the day. Their methods were to strike when least expected, at the weakest point and to melt away before organized retaliation. They gave no quarter and received none from the Grande Armée. As the tide turned against the French, the guerrilleros descended like wolves upon the withdrawing army, and throughout the war the constant threat that they provided behind Napoleon's lines was worth many divisions to Lord Wellington. They proved adept at gathering intelligence and provided a valuable network of spies, producing information which undoubtedly enabled the British to plan their strategy to good effect.

Some years before the Second World War, a Chinese communist leader by the name of Mao Tse-tung was leading irregulars against the Nationalist forces of Chiang Kai-shek and today his doctrines on revolutionary warfare have been shown to be sound. As a result, attempts have been made to carry out guerrilla wars in accordance with these doctrines.

Although he fought against the Japanese invader, it was not his idea in 1937 to end the struggle against the Nationalists. However, at the request of the Comintern, he agreed to co-operate with Chiang and thus weaken Japan, in order to increase Soviet security in the East when Russia was threatened by Hitler in the West. So Mao was able to put over a revolutionary doctrine, disguised as patriotism. Indeed, he benefited greatly from this co-operation. He was able to expand his areas of influence, establish his party organization on a much wider scale and produce a mass movement in support of the communist party. When the Japanese were defeated, he was in a strong position to continue the Civil War.

Mao's principles and maxims for guerrilla action are now the guide to all revolutionaries. Before attempting to understand the weapons and tactics of such warfare, it is important that we should study this doctrine, which is as follows:

(1) The first law of war is to preserve ourselves and destroy the enemy.

(2) 'Our strategy is to pit one against ten, whilst our tactic is to pit ten against one.'

(3) In all operations ensure the collaboration of the local populace.

(4) Retain the initiative by maintaining the offensive, give the enemy no rest. Harass him continuously until he is exhausted, becomes dispersed and can be annihilated. When he advances, withdraw. When he attacks, disperse. When he halts, harass him: use night attacks to disturb his rest and lower his morale. When he seeks to avoid battle, attack him.

(5) Fight only when victory is certain: run away when it is not. 'Avoid the strong, attack the hollow.' Successes are important to boost morale and to capture arms and equipment.

(6) Plan every action carefully on the best local intelligence and co-ordinate your plans with any other units in the area.

(7) Avoid prolonged engagements; always seek a quick decision.

(8) Be alert and act with speed in all phases of combat.

(9) Move secretly and rapidly, mainly by night or in bad weather.

(10) Use every ingenious device to mislead, entice or confuse the enemy. 'Make an uproar in the East and strike in the West.'

(11) To overcome logistic difficulties, live off the country and learn to accept hardships. Avoid reliance on a regular line of supply. Establish and use widely dispersed food and ammunition dumps. Weapons, ammunition, equipment and medical supplies must be obtained from the enemy.

(12) Make every effort to encourage enemy defection. Prisoners should be well treated, and, if possible won over to the cause.

Guerrilla tactics generally follow the principles of Mao, but in studying the insurrections of recent years a common sequence is revealed. Broadly speaking there are five phases, but these are by no means clearly defined and, indeed, tend to be somewhat concurrent.

The preparatory phase caters for the recruiting and organization of the forces. A command network is established and attempts are made to gather sympathy for the cause. This phase is quickly followed by one of terrorism, which today tends to deviate from Mao's doctrine. Terrorists operate alone or in small groups. They aim to secure the support of the people and to gain access to material sources. Weapons may be of the simplest variety in the early stages, but more sophisticated types can be captured from the security forces. As one might guess, their greatest weapon is terror itself. The adherents hope that they may bring about a change in government policy or, indeed, the status of the country by breaking down law and order. They do not intend to do battle with strong security forces, but prefer to tie down police and troops to static guard duties.

There are two simple rules for terrorists in their selection of victims for assassination, kidnapping, damage and disruption. Firstly by killing government officers and sympathizers, damaging government buildings and disrupting communications, they will cause maximum inconvenience to the government. To avoid inconvenience to the local population, whose sympathy they seek, sabotage of places of work should be avoided, to prevent wide-scale unemployment. Secondly, depending on the action of the security forces, it may be more useful at times to attempt to wear down their morale by selecting them as a principal target than to concentrate on the possibly more worthwhile task of disrupting local administration.

Terrorism is cheap and requires relatively few men and arms.

In Cyprus, Grivas is thought to have had only 220 hard core followers and there were about 500 active terrorists in Palestine. Terrorists often live in urban areas rather than remote guerrilla bases and thus their supply requirements are negligible and their supply lines short.

Whilst it is helpful for arms to be provided by a sympathetic nation, many weapons may be obtained or manufactured by the terrorists. Sporting guns and explosive are the basic essentials; thereafter the insurgents depend upon their ingenuity. They may be assisted by attending training courses abroad or even by having foreign advisers attached to their organization.

In small countries lacking concealing terrain or in civilized countries with good roads, terrorism may be the only means available to the rebels. The terrorists can then operate in the cities and towns.

The third phase is often the establishment of guerrilla control in country areas. This is to ensure a measure of safety and concealment for the establishment of supply dumps, training of members and the setting-up of a larger-scale H.Q. to control the campaign. At this point the security forces are becoming the primary target. As the organization expands, it becomes more expensive in men and equipment and outside material aid is usually necessary. If this aid can be cut off by the security forces, there is a danger that the movement may collapse. It is essential that there should be suitable guerrilla country if this phase is to succeed. If the guerrillas are defeated, they must go underground and adopt subversion and terrorism, whilst they build up their strength once again. This may take a considerable time, and in 1970 we may be witnessing the first signs of this happening in Malaya, where communist terrorists seem to be emerging again. If, however, the guerrillas have been successful, they can proceed to the next phase, which should lead to the final overthrow of the established government by open insurrection. The guerrillas may need to operate as an organized army in the field, before this aim can finally be obtained. Outside material aid will be essential, and indeed months of preparation must be devoted to training and the building-up of ammunition supplies and weapons. Whilst this is going on, the security forces must continue to be attacked and harassed whenever possible. It is important that good intelligence should be obtained to permit the planning of a decisive campaign. It is important that the newly formed 'regular' guerrilla forces should meet the enemy on a selected battlefield, where they can match him in strength and weapons. General Giap's Viet-Minh suffered defeats whenever they attacked the French in the Red River delta, which was near to the French main base. Here the air force and artillery could intervene effectively and boats were used to out-manoeuvre Giap's forces. By 1954 Giap had learnt his lesson, and when he attacked at Dien Bien Phu he had spent months concentrating a superior force of artillery. He knew that the French Air Force and armour were too far away to be of real value to the garrison at this critical time.

The danger is that the insurgents may become over-confident and attempt to engage an opponent with greatly superior fire-power on an open field of battle. It is interesting to note that once the Americans moved in large-scale forces to Vietnam, the Viet Cong returned to a large-scale form of guerrilla war.

The Viet Cong used every form of terrorism, subversion, political pressure and guerrilla war against the U.S. forces. Outside aid from North Vietnam took the form of both men and equipment. The situation was unusual because there existed almost open war between the U.S. and North Vietnam, although all the ground fighting took place inside the arbitrary boundaries of South Vietnam. In an effort to persuade the North to end their support of the Viet Cong the U.S. Air Force dropped more bombs than fell on Europe in the Second World War. This aerial bombardment did not close the supply routes. Damaged roads and bridges were quickly repaired by squads of coolies. Locally made houses can be rebuilt in a matter of days and it is found that bombing has little point where targets are so easily repaired. The Vietnam

War brought about the introduction of many new weapons and tactics.

The methods of countering the insurgency are discussed in the next chapter, but let us hope that Vietnam does not turn out to have been a second Spanish Civil War, in that it preceds a world conflict.

# 16. Counter-Insurgency

To overcome the insurgents it is necessary that both a political and a military campaign be waged. Before long-term political, economic and social measures can improve the situation, military action will be needed to restore the control of the legitimate government and bring peace, law and order to the country. Therefore military operations should aim to regain control of the guerrilla-dominated areas, without alienating the support of the people. The insurgents are often dependent upon the support of the people, and without it they must move on. The majority of the population help the guerrillas, because they fear them and without government protection they are at their mercy. In some areas of Vietnam the Viet Cong had been in control for so long that the people regarded American and government forces, who came to liberate them, as aggressors.

The weapons needed to overcome this sort of situation are the radio and the leaflet, used in the general tactics of what we now call psychological warfare. The tactics are well known to those who heard the broadcasts of Winston Churchill and, in Germany, of Goebbels. The importance of winning over the hearts and minds of the people cannot be exaggerated.

It is difficult for the soldier, when seeking the elusive terrorist, to keep to the objectives of the overall campaign. The enemy rarely wears uniform and his appearance is identical to that of the rest of the population. To protect the people from the guerrillas, military operations should be planned to isolate the insurgents. This may entail such measures as food control, curfews and the limitation of all movement. At the same time the guerrillas must be pursued by offensive patrolling, which from time to time will undoubtedly be to the detriment of innocent people. Only the most careful planning and co-operation between military and civil authorities can hope to do all this and win the hearts and minds of the people at the same time.

Although it is necessary to draw the guerrillas to battle and defeat them decisively, it is more important to extend the areas controlled by the government and thus draw support away from the insurgents.

Without good intelligence, the army has little chance of success, but this intelligence may be obtained only by the ceaseless efforts of joint police and military agencies. With accurate intelligence it may be possible to discover sources of supply to the guerrillas that are coming from external supporters. If the route and timing of the deliveries may be ascertained, they can sometimes be prevented. Obviously, on an island, this can be done by active naval patrolling, but in cases where frontiers are dominated by the insurgents and pass through difficult terrain, it may not be so easy to block the supplier. In Vietnam the Ho Chi Minh trail, running through dense forest, is a particular example of this problem.

Insurgent organizations are usually very dependent on the leadership of a small number of dedicated men. The removal of the leaders may have a disproportionate effect on the morale and resistance of the remainder. It may be, therefore, that an important object of counter-insurgency operations should be the elimination of these men.

The troops best suited to this warfare will normally be of the same race as the guerrillas, or similar to it. They may even be defectors. The success of the counter gang in the Kenya Mau Mau campaign is a good example of this. It may even be possible to exploit the allegiance and enlist the aid of small tribes that inhabit remote areas. These minority groups may be able to protect frontiers, prevent guerrilla movement and prove a useful source of intelligence, with only a little support from the regular forces. For political reasons it may be preferable to use the indigenous troops. It will be more difficult for a hostile press to accuse a foreign power of imposing its rule on the country if it is seen that local forces are working for their own deliverance. Initially, the Americans had hoped in Vietnam that they could merely provide advisers, technicians and equipment, to avoid being embroiled in what became a ghastly war. Having been sucked into the conflict on such a scale, the U.S.A. felt that the advantages of their twentieth-century weapon technology and well-trained soldiers could overcome the unsophisticated Viet Cong. In the jungles, mountains and paddy fields of Vietnam, they quickly found that their forces, armed and trained for battle in the nuclear age and in an armoured period, could not subdue fanatical and numerous insurgents. Once again, man had to enter an un-armoured phase because the terrain was against him. Already the armies of the world were seeking a means of fighting a land battle from vehicles that could be independent of the ground and its associated obstacles. Thus came the sky cavalry. This subject is discussed in the following chapter.

The detailed weapons and tactics of guerrilla warfare are the subject of numerous books, and space does not permit a full discussion. The campaigns will be bitter and will be waged without mercy, by men driven forward by fanatical and often ideological doctrines. The task of the counter-revolutionary is to secure the psychological initiative, win their support and allegiance and isolate them permanently from the insurgents. We of the West may be too soft to counter this threat success-fully, but it should be remembered that the acts of terrorism and intimidation committed by the guerrillas will perhaps yield short-term gains, but in time may turn against them. The suc-cess of counter-insurgency may be based on the ruthless sup-pression of the guilty and the firm but courteous treatment of the innocent.

At present guerrilla warfare is the phalanx of communism, but we should not forget that we too have used and may use, in the future, similar warfare.

# 17. The Sky Cavalry – A New Arm

In the Korean War it was seen that the helicopter had an important role in modern war, and in 1961 the U.S. army sent helicopters to increase the mobility of the Vietnamese forces. By 1966, over 1,600 army helicopters were performing a variety of tasks in Vietnam. This included the movement of troops, aerial resupply, reconnaissance and the direction of artillery fire. In addition, these aerial platforms could be used for command and control, casualty evacuation and radio relay facilities.

The disciples of armoured warfare felt that the unarmoured helicopter, flying slowly at low altitudes, would present an easy target for the enemy. However, statistics tend to offer proof of their ability to survive. In 1965, it was said that only one helicopter was lost in every 18,000 sorties. The critics point out that this may be so in counter-insurgency war, but would be very different in conventional war against a sophisticated enemy. As a result, many elaborate experiments and exercises have been conducted to determine if this is true. There have been conflicting results, but it has been established that helicopters are less vulnerable to ground fire than was estimated, and today the armies of the world are increasing the size of their helicopter fleets and training to use them at the very edge of the forward battle area.

When troops were carried forward by helicopter to their objective, there was a dangerous gap in their protection as they flew in and during the actual moment of landing. Like parachutists and glider troops, they were vulnerable until their own ground-based weapons could be brought into action. This was overcome by arming helicopters to escort the transports. Today a heli-borne force appears suddenly at tree-top level. The leading gun-ships rake the objective with a concentrated fire of multiple high-explosive rockets and machine guns. Smoke may also be put down with the suppressive fire, and before enemy weapons can be brought to bear the transports have delivered their troops and withdrawn. Meanwhile the gun-ships continue to support the ground forces, or conventional fighter bombers may be called in by an aerial spotter and directed against ground targets. Naturally all this depends on having air superiority.

To give greater protection to the vital components of the machine and to the crew, considerable efforts have been made to develop the bullet-proof vests (first used by air crews in the Second World War) and light-weight armour plate. If, despite the armed escorts and the armoured protection, the helicopter is disabled, it has the ability to land almost anywhere and therefore a high percentage of these aircraft may be recovered, and, after repair, returned to service.

One of the problems of the helicopter is the enormous amount of maintenance it requires, but experience in Vietnam indicates that, providing there are adequate spare parts and well-trained maintenance personnel, these machines can be kept in action in the battle area for long periods.

All flyers are limited to some extent by bad weather and at night. However, the slow-flying helicopter can operate more readily at times of reduced visibility than its conventional counterpart. If conditions become really impossible, the pilot can switch on a powerful light and land in a small area of flat ground. Nevertheless, successful night operations usually tax the crews to the uttermost of their skill, especially in bad weather.

Heli-borne forces may be resupplied by air. This is expensive in fuel and, compared with a normal ground resupply, cannot carry anything like the same tonnage. However, the vast numbers of men that would be required to guard a ground line of supply are not needed and goods can be delivered more quickly. The air route must, of course, be protected by fighters, both from air and ground attack.

The control of the heli-borne or airmobile assaults demands an airborne command post, and in case of casualties there should be a duplicate system. The helicopter allows the commander a superb view of the ground battle and permits the rapid adjustment of artillery fire and ground attacks by fighters. At the right moment, the commander may be landed and assume his normal ground role.

The tactical effectiveness of the heli-borne force has been firmly demonstrated in Vietnam and, to some extent, by the French in Algeria. Air mobility gives the infantry the great advantage of flexibility and rapid deployment. Such units can operate over vast areas of territory and outflank a ground-based enemy at will.

In the first part of this book, Tom Wintringham has said the blitzkrieg and airborne assaults were aimed at getting behind the enemy's main position. The airmobile division has a similar task, except that in Vietnam there was no front line and the enemy had to be sought out and destroyed in their normally inaccessible positions. Lorries are replaced by transport helicopters, and tanks by gun-ships. If required, artillery may be lifted in by the heavy-lift helicopters, as can light tanks. The units in Vietnam were not designed to replace normal infantry divisions, but, like the parachute forces and the marine commandos, they had a special capability of their own.

There appears to be a number of different categories of helicopter that might come into service. Firstly there is a requirement for a light reconnaissance helicopter – fast, manoeuvrable and having good radio communications. This machine may be used to seek and observe the enemy. The armed helicopter is the tank of the sky, able to deliver a powerful blow and carry a useful range of weapons. It should be compact and its vital parts protected against the inevitable ground fire. Lastly there are the transports. These will be required to lift men, equipment, supplies and weapons for long distances at a reasonable speed. They may need some light weapons for their protection and they will probably be categorized by their load-carrying ability.

In Wintringham's second unarmoured period (A.D. 378–774), cavalry became the main arm that won battles; it was usually a fairly light cavalry, not fully armoured. They fought with missiles more than by shock action at close quarters. The period ended with the appearance of the heavy armoured knight, who could not be destroyed by the lightly armed horseman. It was not until the longbow could penetrate armour that the value of the knight declined. Now history repeats itself, and today anti-tank guided missiles capable of destroying tanks at ranges in excess of 2,000 metres may be fired by helicopter. The slow-moving tank is a better target than its nimble adversary. The tank can of course conceal itself more readily and no doubt there are now weapons and tactics that may be used to counteract the helicopter threat. The introduction of airmobile forces will, I predict, have a lasting effect on our future weapons and tactics. Armies may now move large numbers of men at speeds of around 100 m.p.h. They may fly over the difficult terrains and the man-made obstacles, over water and even in conditions of moderately bad weather. We ignore these new developments at our peril.

## 18. The East Glows Red

In the early morning of the 6 August 1945, Colonel Paul W. Tibbets piloted his Super Fortress in a cloudless sky towards Hiroshima. His target was one of the largest towns of Imperial Japan. Here, amongst the arsenals, factories and oil refineries, lived 343,000 civilians and over 150,000 soldiers of the garrison.

Hiroshima had been selected as the first military target to feel the effects of nuclear war. As the Super Fortress turned away, the world's first atomic bomb descended on its parachute towards the city. A Japanese journalist describes the scene:

Suddenly a glaring whitish pinkish light appeared in the sky accompanied by an unnatural tremor which was followed almost immediately by a wave of suffocating heat and a wind which swept away everything in its path. Within a few seconds the thousands of people in the streets and the gardens in the centre of the town were scorched by a wave of searing heat. Many were killed instantly, others lay writhing on the ground screaming in agony from the intolerable pain of their burns. Everything standing upright in the way of the blast – walls, houses, factories, and other buildings – was annihilated and the debris spun round in a whirlwind and was carried up into the air. Trams were picked up and tossed aside as though they had neither weight nor solidity. Trains were flung off the rails as though they were toys. Horses, dogs and cattle suffered the same fate as human beings. Every living thing was petrified in an attitude of indescribable suffering. Even the vegetation did not escape. Trees went up in flames, the rice plants lost their greenness, the grass burned on the ground like dry straw. Beyond the zone of utter death in which nothing remained alive houses collapsed in a whirl of beams, bricks and girders. Up to about three miles from the centre of the explosion lightly-built houses were flattened as though they had been built of cardboard. Those who were inside were either killed or wounded. Those who managed to extricate themselves by some miracle found themselves surrounded by a ring of fire. And the few who succeeded in making their way to safety generally died twenty or thirty days later from the delayed effects of the deadly gamma-rays. Some of the reinforced concrete or stone buildings remained standing, but their interiors were completely gutted by the blast.

About half an hour after the explosion, whilst the sky all around Hiroshima was still cloudless, a fine rain began to fall on the town and went on for about five minutes. It was caused by the sudden rise of over-heated air to a great height, where it condensed and fell back as rain. Then a violent wind rose and the fires extended with terrible rapidity, because most Japanese houses are built only of timber and straw.

By the evening the fire began to die down and then it went out. There was nothing left to burn. Hiroshima had ceased to exist.*

At Hiroshima approximately 78,000 died at once and 10,000 were reported missing. There were 37,000 injured and many died or suffered later from the gamma-rays. Before the Japanese could be persuaded to surrender, a second bomb had been dropped at Nagasaki and another 40,000 killed.

The decision to produce this terrible weapon in the U.S.A. had been taken in the early part of the war, when it was known that German scientists were working to develop a similar device. Fortunately for the Western allies, they won the atomic race and the war in Europe ended before Hitler could perfect this means of mass destruction.

There is little doubt that had he done so he would have used it. The moral revulsion that followed the atom bombing of Japan is still felt to this day, but war is an ugly business and there is little difference between death by bomb, bullet, fire, gas, flame or nuclear explosion. The bombing of Germany and in particular the fire storm created at Hamburg was every bit as horrible to the individual victim as the death dealt out that August morning in Japan.

*Quoted by Desmond Flower and James Reeves, *The War, 1939–1945* (Cassell, 1960), pp. 1031 and 1032, from Marcel Junod, *Warrior without Weapons* (Cape, 1951).

The Second World War ended in a crescendo of violence, with suicide pilots of the Japanese Air Force driving their explosive-laden aircraft into the decks of the Allies' ships. The Russian army had arrived in Berlin to mete out a terrible revenge for the German atrocities on the Eastern front. The gates of Belsen were opened and the surviving prisoners hobbled out of Changi Jail. The whole world lamented the catastrophic events of the past six years and vowed for at least the second time that war should be no more.

In spite of all the good intentions the bomb was here to stay and we had to learn to live with it. It undoubtedly caused a complete reappraisal in the development of weapons and tactics. Naturally the first thought was for protection, and as the weapon rapidly grew larger and more powerful a sense of futility began to reign. This was, to a great extent, prompted by the growing economic burden of weapon development. It became obvious that only a very wealthy nation could afford an effective army in the Nuclear Age. However, the tactician's aim is to make the best use of the men and weapons at his disposal. To achieve this under the threat of mass destruction called for greater mobility and protection. Thus, since 1945, all armies have developed fast-moving armoured formations and self-sufficient all-arms groups.

In any nuclear conflict the problem of maintaining the fighting man's morale will be aggravated by the threat of severe damage to rear administrative areas and to the homeland.

A defensive attitude reappeared and, in the face of Soviet threats, alliances, such as N.A.T.O., sprang up. Vast sums of money were spent on the development and production of nuclear weapons, methods of delivery and air defence. While the costly stalemate continued the tactician was somewhat belittled by the strategist and it was the Korean War that reawakened the idea of conventional non-nuclear fighting.

Hardly had the dust of war settled, before ominous rumblings were heard again in the East. At 0400 hours on Sunday, 25 June 1950, 90,000 North Koreans with over 100 tanks invaded South Korea. In spite of warnings that this might happen, it seems that both the South Koreans and their American friends were taken by surprise. It has been described as a 'limited war', although on the communist side there were few limits. It was the first war in which a large-scale Chinese army clashed with a Western force and the Chinese communist soldier, although hardy and self-sufficient, was found to be no superman. At the very moment when the communists feared that American patience was exhausted and nuclear weapons might be used, they sued for peace. During the months of negotiation that followed, the Chinese and North Koreans succeeded in recovering very nearly all their lost territory.

It was a war fought in a barren, inhospitable land by a multi-national army of United Nations forces. Fighting largely with the weapons and tactics of the Second World War, the pre-dominantly European/American troops soon found they had to modify many of their ideas. They found that, although the Chinese and North Korean forces were often capable of large-scale movement by night and good camouflage, they were generally not good in mobile warfare. But it is true that some elementary mistakes were made by the U.N. forces and that by sticking to the roads they were often outflanked by the enemy moving along ridges and hilltops.

The communist propaganda machine was so good that even today in some Western quarters it is believed that the Chinese forces were found to be invincible and victorious in Korea. This is completely untrue and indeed the sheer fire-power of the U.N. army overcame a million of the best troops China could produce.

The communist armies worked under the close supervision of their political commissars. Their standard of training was low and there was little scope for initiative. It is thought that the so-called mass tactics were merely the results of badly trained soldiers bunching together as they approached the objective. In the West, recruits are constantly warned to keep five yards apart at all times, but lack of communications may have dictated the need for the communist troops to keep close together. On the other hand, it may be argued that the communists simply tried to concentrate their forces at a point when they attacked the U.N. line, and by so doing hoped to drive a wedge through the defensive positions. In the face of these tactics, U.N. forces found themselves being overrun by a human sea, which could not be stemmed even by the most terrible casualties. The U.N. found that their weapons were out of date and they needed rapid-firing, shorter-range rifles, small, light maiming mines and more ammunition well forward. The real danger of infiltrating Fifth Columnists was felt in the rear areas. As far as static defence went, there were few lessons that had not been learnt as long ago as the 1914–18 War.

The air war in Korea was fought by newer and more powerful aircraft than had been used in the Second World War. As in the days of Douhet, there was a school of thought believing that air power could end the war almost by itself. However, although the U.S. air forces had complete air superiority and were able to bomb installations at will, supplies and reinforcements continued to reach the enemy front.

There was a reluctance to attack strategic targets, such as the important dams on the Yalu river, and if these had been destroyed it would certainly have made the enemy's supply problem more difficult. The U.N. always hoped that they would bring the communists to the conference table and thus they were reluctant to destroy facilities which it would be difficult and costly to replace after the war. In the air, the helicopter became an important military vehicle and it was its success in Korea that led the U.S.A. to concentrate on its development, later to prove so important in South-East Asia.

## 19. The Armour Develops

As armoured mobile armies developed there came the problem of command and control. Since 1918 a considerable advance had been made in military signals, and the armies of the sixties were equipped with a scale of radio sets hitherto unknown. Today electronic gadgetry is still improving and computers are leading the advance.

In order to defeat an armoured aggressor the defence had to be more mobile and based on the skilful use of ground and obstacles. If the obstacles were to be effective they needed to be covered by observed fire and thus once again military scientists were asked to find a way of increasing the range and accuracy of guns.

After the criticism British armour had received in the Second World War there was much discussion on the requirements for our future tanks. These costly beasts were needed in large numbers during the conflict and with the coming of peace and financial stringency the army had to be sure that it got good value for money. For many years the importance of the tank was considered to be its armour protection. Critics may argue that a tank's value is reduced when it can be penetrated by anti-tank weapons. However, especially in the Nuclear Age, we see that nothing is invulnerable. Today designers attempt to produce a tank that is more mobile and thus has the ability to carry a heavy weapon quickly about the battlefield. There must be a balance between fire-power, mobility and protection. The tank's most difficult target is another tank, and thus the ability to destroy enemy armour is of primary consideration in tank design. Armoured formations must be able to counter large invading forces of enemy armoured fighting vehicles, for these will undoubtedly form the spearhead of aggression in many areas of the world. In particular the North German plain is eminently suitable for tank warfare.

Our tank guns in 1940 were far too light and proved ineffective against the German armour. A tank gun must first of all be able to hit the target and for this much depends on the skill and training of the crew and the accurate estimation of range. Mainly owing to the high standard of training, the Israeli Centurion tanks achieved remarkable results in the Six-Day War of 1967. They were faced by Egyptian units equipped with relatively modern Russian armour which was dug in, in prepared defensive positions. The Israelis were able to score hits on the enemy tanks while they themselves were well outside the range of the opposing tank guns.

Modern guns firing shells with a very high velocity and having a flat trajectory give a higher probability of a hit than those with lower-velocity high-trajectory ammunition. However, high-velocity guns have high recoil and require a heavier mounting. The high-velocity round is a solid shot of extremely strong metal which is designed to punch a hole through the enemy's armour plate. High explosive is used in the slower speed ammunition and by one means or another blasts a hole in the target. The nub of this problem is that it is easier to hit a target if you can fire straight at it than if you have to lob your shell at it. When firing direct, an error in range estimation matters little, but when lobbing your low-velocity shell an error of fifty metres in estimating the range can mean a miss. At above 1,000 metres in the chaos of battle it is easy to make such a small error. Even with modern automatic range-finders such errors may occur and the high-velocity gun is still the most accurate.

For some years after the Second World War it was thought that 2,000 metres would be the maximum range in tank battles, but in order to outrange the opposition this distance has now risen to over 3,000 metres. The lesson of the Six-Day War has given impetus to this requirement, although there are still doubts about the frequency with which tanks will be able to acquire targets at such ranges.

In order to kill the enemy armour at increasing ranges there has been a trend since the war for tank guns to get larger. However, the increase in calibre of high-velocity guns has caused much heavier recoil loads. These eventually mean heavier tanks, and if more armour plate is added at the same time the vehicle becomes less mobile and may not be able to use existing bridges.

Thus the knight in his armour must be replaced by something equally effective but lighter. In the 1950s the French Army decided to use lighter, more mobile tanks firing low-velocity projectiles that caused less recoil and depended on 'shaped-charge' shells. These penetrate armour plate by focusing the blast of the high explosive. Therefore the French could produce the A.M.X. 30, which weighed only 32 tons, as compared with the latest British tank, the Chieftain, which was over 50 tons. The A.M.X. 30 is fitted with an optical range-finger to assist the gunner, but it is not as reliable as the laser range-finders now under trial for the tanks of the future.

The fitting of guided missiles (G.M.) to armoured fighting vehicles is already commonplace in the world's armies. Owing to their numerous limitations, the first-generation G.M. do not represent a practical alternative to the tank gun.

The small G.M.s such as the Vigilant need only a little space for its launcher and control equipment, but the missiles are complex, costly and bulky. In order to carry several G.M.s inside a tank extra space must be provided. They are too vulnerable to be stowed outside. Tanks may have to use their guns to engage various types of target (e.g. men, vehicles, buildings, tanks) and carry a variety of ammunition for this purpose. It would not be a straightforward matter to switch warheads on the missiles and they have no 'canister'* capability to use against massed infantry. The G.M. has a slow flight, but to enable the operator to get control of the missile a few seconds'

*A thin metal round or canister filled with metal balls that enables a gun to be used on massed infantry with a shotgun effect.

flight time is required. Thus the weapon has a minimum range inside which it cannot be used. Reloading is a slower process than with a gun and each missile costs many thousands of pounds.

Tactically the current anti-tank G.M. is suited to use by infantry long-range anti-tank defence in open country, when the target is unlikely to disappear behind cover while the missile is in flight.

However, anti-tank missiles of the seventies are smaller, faster and have semi-automatic guidance which makes them more independent of their human controllers. In the case of one new system the missile launcher is also capable of firing conventional unguided missiles, which greatly increase the versatility of the weapon.

It is said that such G.M. systems offer, in theory, a high hit probability and will mean that it is possible to build lighter, more mobile tanks. However there are at present technical problems and their reliability has yet to be proved on the battlefield. There is also the risk that, if we concentrate on the use of shaped-charge projectiles, the enemy need only armour their tanks against this form of attack. It is much simpler to design a tank's armour if it is known that only one form of attack will be used against it. On the other hand G.M.s can be made to carry much larger explosive loads and it is difficult to imagine that in the end a tank could be immune to these weapons. At present G.M.s do not offer a completely satisfactory alternative to high-velocity guns, which are capable of further development especially in the field of higher muzzle velocities. To hit the target is not in itself enough: the armour must be penetrated and the internal damage must be lethal.

Many attempts are being made to find a way of maintaining the fire-power, mobility and protection of the tank and yet reduce its weight. So far there has been little success in this direction, although aluminium alloy armour is used on Armoured Personnel Carriers (A.P.C.s) and air-transportable tanks. Further saving may be possible as more compact engines are developed, but we cannot afford to reduce either the speed or the armour. Strangely it is very difficult to increase the cross-country speed of tanks. Anyone who has ever ridden in one of these monsters at high speed will appreciate that there is a limit to the roughness of a ride that the human frame can tolerate. Although improved suspensions may help, it is unlikely that fighting machines that must move in contact with the ground can go any faster. The problems of weight largely govern the mobility of tanks, but although we may reduce their weight we are limited by the type of gun available. A possible solution is the use of recoilless* guns on an armoured chasis. These weapons are light and impose no firing load on the vehicle so that a low-velocity explosive shell with good armour-penetrating capabilities may even be used from a Land Rover. Unfortunately the range of recoilless guns is limited and they are rarely effective beyond 800 metres. Their ammunition is heavy and there is a great back blast when they fire which makes them extremely difficult to conceal. However, as an armament for use at close range by a light vehicle, they have much to recommend them. It may be that such guns would be suitable for use on hovercraft-type fighting vehicles, where any form of recoil would be unacceptable.

Since the Second World War great improvements have been made in night viewing equipment. The Soviet Army spend a high proportion of their time training to fight and move under the cover of darkness. Their tanks and many other vehicles are fitted with infra-red devices which enable the drivers to see a short distance at night. Their armour also carries infra-red searchlights which may be used to illuminate targets for the tank gun. The same searchlights may be used to produce a dazzling beam of white light and one might think that this device would immediately attract fire from the opposition, but it is very difficult to hit a searchlight that merely comes on for a few seconds at a time on a tank which may move in between illuminations. The use of infra-red is almost as dangerous, for

*A simple, ultra-light gun with an open breach. When fired a proportion of the propellent is used in a backward blast and this equalises the recoil. Thus the gun remains still.

the enemy has only to be using a passive infra-red viewer and he will immediately detect all sources of this light, including hot exhausts as well as infra-red lamps. There is one other problem in this field and that is that missiles can be made to home on infra-red sources.

In spite of its armoured protection and its night viewing devices, the tank is extremely vulnerable after dark and needs protection from enemy tank hunters.

Today there is controversy over the tactical use of A.P.C.s and the best method of anti-A.P.C. defence. To use an anti-tank gun to destroy a lightly armoured troop carrier is to use a sledgehammer to crack a nut. Vital anti-tank ammunition must be conserved for the 'elephants', but, at present, there is a lack of less expensive missiles only sufficiently strong to pierce the 'shield'.

All sophisticated modern armies use a tracked or wheeled A.P.C. to carry infantry on the battlefield. The tactical handling of these vehicles in the assault are an important feature of modern warfare.

Two distinct doctrines of tactical handling have developed, one sponsored by Britain and the other by West Germany. In the British view the vehicle is used solely as an armoured means of transport to the objective; but the Germans believe in actually fighting from an A.P.C., if necessary. The British school of thought holds that A.P.C.s are essentially transport vehicles and assumes that their principal function is to bring infantry forward for dismounted action. The troops are carried in a relatively high-sided 'box', mounting a machine-gun as its sole armament. The majority of the crew cannot see the ground over which they are moving, but are briefed by the section commander over an internal loudspeaker system.

When tanks are available to give close support, they usually lead the A.P.C.s onto the objective. Both tanks and A.P.C.s should aim to arrive at the dismounting position at the same moment. The infantry will normally dismount on the objective, if the ground and the defender's fire permit. When this is impossible, dismounting takes place in an area unlikely to be mined or selected as an enemy artillery target. The infantry then works its way forward onto the objective using cover, and the tactics of fire and movement. Once the sections have dismounted the A.P.C.s will either remain on the objective to give covering fire with their machine-guns or withdraw to a rendezvous under cover. The dismounting position may be pre-planned or confirmed by the commander on the radio as the attack begins.

Reserves may remain aboard their A.P.C.s, where they are protected and in contact so that they may be moved rapidly to any critical point in the assault.

The British approach increases the overall mobility of the infantry, but it implies that it is to do all its fighting dismounted. It does not consider the possibility of the use of personal weapons from the vehicle, although all the normal support weapons may be fired from the A.P.C.

A turret-mounted cannon is being developed for anti-A.P.C. action. The cannon will give a substantial increase to the platoon fire-power. However, at present no attempt is being made to introduce loopholes or gunports for the use of personal weapons.

The U.S. army have, in general terms, followed the same A.P.C. tactics as the British. Their vehicle is very similar to the British model, but now experience in Vietnam has caused them to increase the fire-power of the A.P.C., supplementing the single machine-gun with other light automatic weapons. The introduction of gunports for personal arms and the additional use of 40-mm. grenade launchers, flame-throwers and heavier-calibre support weapons from their A.P.C. marks a change in policy towards the creation of an Infantry Fighting Vehicle (I.F.V.): that is to say a vehicle from which infantry actually fight.

The West German school of thought considers that A.P.C.s should be vehicles from which the infantry can fight on the move, dismounting only when absolutely necessary. They have developed A.P.C.s to suit this doctrine and have paid great attention to the lowering of the silhouette and the sloping of the

armour plate in order to make the vehicle A.P.C. less vulnerable. All the occupants can use their personal weapons and the
vehicle has a turret-mounted automatic cannon. All this enables the West German A.P.C. to act as an I.F.V. and to fight
its way further forward before dismounting becomes necessary.
By comparison, some other A.P.C.s are more vulnerable because of their higher silhouette and the inability of their crews
to use their personal weapons for close-range defence.

The 'tin box' A.P.C.s have their own virtues, but there is
much to be said for the West German approach. Their emphasis on the fighting role of armoured infantry vehicles is
backed by considerable practical experience. Shortages of tanks
and other equipment during the later stages of the Second
World War made it necessary for them to make the most of
what they had and, driven to it by necessity, they achieved
considerable success with mobile battle groups equipped largely
with armoured infantry vehicles of I.F.V. type. This was particularly the case during mobile operations on the Eastern
Front, in Russia, where I.F.V.s were commonly used for
mounted mobile combat. In spite of this experience, the German A.P.C. suffers from a lack of amphibious capability,
which may well be a serious disadvantage in a fast-moving
campaign.

The Soviet army also favours the I.F.V. concept, but
appears to be more inclined to dismount than the West Germans. When facing a strong defence the infantry will dismount 500–1,000 metres from the enemy and advance about
100 metres behind the tanks, supported by the fire of A.P.C.
machine-guns and other covering fire. In bad going the infantry will lead. Like the West Germans, the Russians argue
that the section must be supported by the fire of its A.P.C. and
as a result their vehicles carry a substantial armament.

As leading exponents of mobile warfare, the Russians are
extremely conscious of the need for A.P.C.s to be amphibious
and their vehicles are propelled by water jet propulsion units in
order to swim fast rivers. The concentration on the boat-
shaping of A.P.C.s has caused them to avoid rear doors and the

consequential water-proofing problems; thus exit is over the sides and the vehicles are often open-topped. This greatly reduces the protection afforded to the crew, but it is understood that overhead protection can be fitted. The open-topped design does however permit the occupants to see exactly what is happening and enables them to use their weapons with an almost unrestricted traverse. To attain a good standard of accuracy with their personal weapons, Soviet soldiers are subjected to realistic training, involving A.P.C. mock-ups that rock and sway whilst the occupants engage targets.

The French seem to have adopted a compromise and although their A.P.C. is designed as an I.F.V. they normally fight dismounted. They believe that in dismounting speed is essential and they train to do this while the vehicle is moving. If necessary the French armoured infantry will fight from the I.F.V., but this is only likely to be against light resistance. The A.P.C. machine-gun may be used to support dismounted infantry, but the primary role of the vehicle is to permit rapid manoeuvre.

On occasions the French have experimented with small, low-silhouette 'tankettes', manned by a two-man crew. These would appear to be a form of I.F.V., but the only influence they have had is to reduce the silhouette of French A.P.C.s.

Dismounting to fight has its advantages. If an A.P.C. drives onto the objective and is hit by an anti-tank weapon, there is a good chance that the entire section may become casualties, but by dismounting out of range this may be avoided. The infantry can work its way through the enemy position in the traditional manner, unaffected by anti-tank fire and able to deal with such weapons as oppose the tanks. The additional fire of the A.P.C. machine-gun provides valuable support for the section. Infantry on foot is less likely to miss small pockets of enemy than those in A.P.C.s. Once the objective is captured there is an immediate threat of counter-attack and infantry in A.P.C.s is vulnerable to long-range anti-A.P.C. fire, whereas soldiers on foot can quickly take up a concealed defensive position to hold ground and can easily advance over almost any terrain.

Under the British doctrine there is a disadvantage that the infantry arrive without having seen the ground and is suddenly pitched into battle. It is extremely vulnerable to ambush in close country and simple obstacles such as a concealed wire fence will hamper the assault. The action of halting to dismount may destroy the vital momentum necessary for surprise and shock action. Those few seconds may be sufficient for the enemy to recover and open fire with his own infantry weapons.

Mounted fighting also has advantages. It should be accepted that in the face of a strong enemy position, well equipped with anti-A.P.C. weapons, it would be foolhardy to use the mounted attack. However, there are times when such an attack may have many advantages. Providing the crew can see and use its personal weapons it will arrive on the objective at high speed, presenting a difficult target for close-range anti-A.P.C. weapons and immune to small-arms fire. By using automatic weapons and the facility of the A.P.C. to carry an abundance of ammunition it can sweep the objective with heavy and reasonably accurate fire. Being in constant radio contact the assault is extremely flexible and can maintain momentum by overcoming wire and simple obstacles. In the reorganization the A.P.C.s own cannon may be valuable for defence against a counter attack.

The A.P.C. is generally a large target and is vulnerable to mines, molotov cocktails and unsophisticated anti-tank weapons. Therefore a mounted assault on even an ill-equipped, but courageous enemy may be more hazardous than a dismounted attack. The gunports and in the worst case an open top reduce the protection to the crew, and there is always a danger that the occupants may be reluctant to leave the safety of their A.P.C. and tend to use it as a light tank, a task for which it is not designed. Mounted attack, as with cavalry in 1918, is only valid if there is no obstacle that can stop a vehicle close to the objective where the defenders can shoot it up.

With the variety of world conflicts it is difficult to favour any one particular technique for all events. In view of the diverse nature of modern warfare it is dangerous to insist that infantry should always fight dismounted and it may be wise to adopt a more flexible approach. Tactics can be modified very quickly, but the design and production of an A.P.C. to suit them will take years.

To accommodate changing tactical concepts an A.P.C. should possess a number of features. It should have the capability to support the section if it dismounts and if possible provide anti-A.P.C. fire. Heavy support weapons are required by the infantry and it is desirable that these should be capable of being fired from the vehicle. The addition of extra weapons may raise the overall height of the A.P.C. and thus the original height should be kept as low as possible to aid concealment and reduce vulnerability.

In the interests of crew orientation, morale and fire-power some provision should be made for gunports. Some form of airtight lid should be fitted to the port for use in cases of nuclear or gas attack, when an air-conditioning plant should be used. The use of numerous periscopes seems an expensive means of solving a simple problem. The design should also permit the occupants to use their weapons on the move and to have all-round observation.

It is important that the cross-country performance should match that of the battle tank if the two arms are to act together. Wheeled A.P.C.s are faster on roads, but for cross-country performance it is considered that tracks will be necessary, although it must be accepted that wheels have a better life, are less vulnerable and are quieter than tracks. In mobile war an A.P.C. should be fully amphibious, otherwise it will be dependent on the limited engineer river crossing facilities already required for the armour.

The smallest fighting sub-unit in the British Army is the section and it is not considered prudent to split this into two or more A.P.C.s. Any vehicle required to carry more than a section is likely to have reduced protection and an increased payload requiring a more powerful engine and greater track width to give it good mobility. This in turn creates a larger and more vulnerable target. However the A.P.C. should be large enough to carry sufficient supplies for the section to last two days without replenishment. This figure probably represents the most one could reasonably expect a section to have with them at any one time.

The A.P.C. should protect its occupants from small-arms fire and shell splinters. The probable effects of nuclear weapons or gas must also be minimized. It would seem that a rear door is preferable for dismounting. It may be faster for the section to dismount over the side, but in that case overhead protection would be reduced.

The concept that mechanized infantry should fight only in a dismounted role is open to doubt. There is evidence to suggest that we of the West need to revise our A.P.C. tactical doctrine and consider mounted action on the objective. Accepting that there are times when a strong enemy position will require infantry to assault dismounted, there are other times when it would be an advantage to fight from the vehicle. This would certainly be the case if the enemy were surprised, or shaken by some recent and unpleasant experience!

Supposing this to be the case, a new A.P.C. is required that will permit the British infantry section to fight from it.

# 20. A.B.C. Warfare

The horrors of Atomic, Biological and Chemical (A.B.C.) Warfare are no longer in the realms of fantasy and science fiction. Since the First World War gas has only been used in a very few cases, and then usually against completely unprotected and relatively primitive armies. It was used by the Italians in Ethiopia and, more recently, by the Egyptians in Yemen. It has also been used extensively in its milder forms for riot control and for flushing underground hideouts in Vietnam.

The war gases form the basis of chemical warfare although defoliants are also in this group and have been used in Vietnam. The gases have various effects. There are those which cause temporary discomfort and others which may be fatal within a few seconds of reaching the victim. Some are of a persistent or lasting nature and can cause casualties a long time after they have been delivered. Non-persistent gas will drift away relatively quickly. The non-lethal group includes tear gas and irritants. Tear gas has an immediate effect on the eyes, causing smarting and floods of tears. The symptoms quickly die away when the attack is over and there is little danger of lasting effects from this weapon. The irritant gas is more unpleasant and consists of compounds which cause irritation and acute pain in the nose, throat and breathing passages. Violent vomiting and a feeling of depression are other symptoms. A mixture of tear and irritant gas is an especially effective weapon for dispersing crowds, but if used correctly under favourable conditions tear gas is often enough.

Another group may be called paralysing gas and although this has little effect in the open, if used in a confined space, low concentrations may cause giddiness and headaches, but a complete recovery is usually fairly quick. There is a danger that if a high concentration occurs, the victim will suffer paralysis of the respiratory centre of the brain and death will follow rapidly.

Gases of a lethal nature include arsine, which, when absorbed through the lungs, can eventually cause anaemia and later death. Choking gas causes severe inflamation of the lungs, which leads on to their filling with body fluid, thus drowning the sufferer. The mustard gas of the First World War was a liquid, giving off an invisible vapour that smelt strongly of onions. It is extremely persistent and like other blister gases, will attack any part of the body it reaches and will penetrate normal clothing. The liquid can cause permanent blindness and sets up terrible blisters. Even the vapour attacks the skin and can cause temporary blindness. Should the gas be inhaled, bronchial pneumonia may result.

The last group to be considered is the nerve gases. They are highly toxic and act speedily. Being odourless and invisible, the only warning of their presence is through a special device carried or worn for the purpose. The vapour may be inhaled or the liquid absorbed through any part of the body; either way, the results are the same. The victim's pupils contract, his vision is dimmed and he finds it difficult to focus. Within seconds severe earache occurs, and is followed by a running of the mouth and nose. The chest becomes tight and breathing difficult, nausea and vomiting begin, followed by a twitching of the muscles, convulsions and, almost as an anti-climax, death!

There are, of course, protective measures that can be taken and protective clothing that may be worn. Ships, tanks and vehicles may be fitted with air-conditioning plants and filter systems to protect their crews. Decontamination, by washing down with strong detergents, is now a standard drill in many armies. Chemical warfare agents may be used on the battlefield to kill or harass the enemy and, in the case of persistent gas, to form barriers and obstacles. Weather conditions may have a considerable effect on the use of these weapons. Soil that has been heated by the sun will, in turn, heat the air above it, thus causing convection. At night however, when the ground is cooler, the air is likely to remain still. Thus, on a warm summer's day, any concentration of gas that is built up will soon rise well above ground level. As in the case of fog, the gas will tend to stay at ground level during the night. Wind can blow away the vapour, but a gentle breeze may cause an increased rate of evaporation of a liquid agent. Thus it may be possible to build up a moving cloud of concentrated gas. Warm ground will increase the rate of evaporation, but low temperatures may freeze a liquid agent, so that it becomes dangerous at some later time. Porous soil may even absorb the gas and rain, of course, will tend to wash it away.

To build up a concentration of gas sufficient for the task may require a large number of shells or rockets. Commanders must consider whether the likely effects warrant the expenditure of ammunition and the programming of the delivery means. Should the wind change, there is always the risk of one's own troops suffering. Gas cannot destroy tanks or buildings, but the effect that its use will have on badly trained or undisciplined soldiers may well be out of all proportion to the actual casualties produced.

Biological warfare is rather different. Unlike chemical agents, biological weapons may be produced cheaply and easily in any good-class laboratory. Even small nations have the ability to manufacture germ weapons. Successful attacks with biological agents can cause temporary disablement, serious injuries or death as the result of an artificial epidemic. It is not easy to keep the germs alive for a long time and they are greatly affected by temperature. They do not penetrate clothing in the same way as the chemical agent and, by and large, gas masks, protective suits and immunization provide a defence.

The unpredictable nature of biological and chemical weapons and the problems of delivery do not make them a popular weapon. Although we prepared for the onslaught in 1939, the fact that it did not come may have been due more to our second strike capability than to Hitler's humanity. Inhumane and horrible as they may be, they are just another weapon and it would be unwise for future generations to neglect the study of this subject or fail to take precautions for their protection.

To understand the tactics of nuclear war one must have a basic knowledge of the effects of a nuclear explosion. Although we think of a war fought with such weapons as being global suicide, there are in fact small atomic shells which may be fired from a gun or delivered by a short-range rocket or even pre-placed and fired by remote control. It is quite possible to have tactical nuclear weapons although their use is likely to lead to swift escalation.

In the firing of a nuclear device there are three major immediate effects. Firstly there is a blinding flash of light, which is accompanied almost at once by an intense radiation of heat. This is followed by a very powerful wave of blast. And finally there is nuclear radiation. Under some circumstances there may be the hazard of residual radiation and fall-out. In general it would seem that the greatest amount of casualties are produced by the heat and blast, but much depends on where the weapon explodes in relation to the surface of the ground. If it explodes at ground level, a considerable amount of earth will be vapourized and taken up into the characteristic ball of fire, where it will be contaminated with radioactive material, so that when it falls back to earth it will be extremely dangerous. Because part of the fireball is obscured, the heat radiation will be reduced. Much of the blast will go to produce a ground shock, which may be used to destroy underground targets. Such targets would be protected from air blasts by the earth above them. In the case of sub-surface explosions, the air blast is greatly reduced.

A deep underground burst will form a 'bubble' under the surface, probably causing the terrain to bulge. The exact effect depends on how deep below the surface the weapon has exploded and the geology of the target area. Up to a certain point in depth, the weapon will vent, casting up a massive amount of radioactive spoil. Once again this will cause hazardous fall-out. There will be little shock wave and, of course, no emission of heat or light. Clearly a nuclear weapon bursting above the ground at a low altitude will give the most destructive effect on surface targets.

Today, armies are trained to take simple protective measures under nuclear attack. Tests with live bombs and dummies have shown that the trained soldier has a reasonable chance of survival if he takes these precautions. The unprepared civilian population has far less chance.

No sane person wants nuclear war, but it would be foolish for us not to shape our tactics so that, in the event of the worst happening, we can adapt them to suit the situation.

## 21. On to Armageddon

Since 1945 the Western Allies have watched Soviet Russia with grave suspicion. There have been periods of tension and times when the communists seemed eager for an end to the Cold War; times when they have seen the defensive measures of the West as aggressive and provocative. Whilst the attentions of both camps are centred on earthly differences there is little chance that the situation will change radically. I foresee that until we bump into an extra-terrestrial enemy we shall continue quarrelling with one another and that wars between human beings will go on until we are compelled to band together to defend our race. The writings of Mr H. G. Wells may indeed come true!

One hope for peace lies in the maintaining of the East–West balance of power. Neither side dare start a war because of the certain retaliation and destruction that will be brought upon it. The bomb has outgrown the world and no sensible person wishes to start a conflict which would yield the victor a ruined earth with a polluted atmosphere. Having said that, it is worth remembering that there is always a chance that a nuclear war might not be the end of the entire world. Although great areas could be devastated, humans might well continue to exist in an untouched region.

However, some madman may always pull the nuclear trigger or some policy may suddenly have completely unforeseen results which may lead to escalation and nuclear war. There is no guarantee that the right balance of power will prevent war. Certain states may still want to impose their will upon weaker neighbours and they will attempt to get their way through conventional warfare. Some nations may adopt this course believing that they are able to play a game of brinkmanship and not qualify for nuclear retribution.

It may be that the majority of conflicts from now onwards will be between the less developed nations and it is to be hoped that U.N.O. would eventually be able to intervene and at least bring about a cease fire. However U.N.O. is a slow-moving beast and it is likely that a world power will become involved in peace-keeping operations in the early stages, until such time as a U.N. force can assemble.

The dangers of escalation and nuclear retaliation are likely to mean that wars will be of a limited nature or simply a guerrilla campaign supported by an outside nation and countered by the legitimate Government aided by a friendly state. But there is still the possibility of conventional war of some magnitude, such as the Arab–Israeli campaign.

In spite of this, the shadow of nuclear conflict hangs over us and all warfare is now influenced by the threat. The momentum of the war may well be such that hostilities would cease a few weeks or even days from the outset. By this time there could be the most terrible damage to the contestants and their neighbours. Nuclear weapon stocks could be exhausted and terrain so disfigured that armies would grind to a halt.

One can imagine two great powers, their backs broken, still trying desperately to strike at each other, each attempt becoming more feeble, whilst in the ashes new warlords arise representing vested interests and continue a struggle by conventional means. Their objectives may have nothing to do with the aims of the original hostility.

In past wars there are always some soldiers who, having survived their first action, grow to almost enjoy the contest. They may be able to shut their minds to the misery and suffering, but in nuclear battle men will rapidly become sickened and shocked and their morale will sink quickly to a low ebb. The problem of keeping troops fighting and retaining their will to fight cannot be underestimated. Commanders will find that they must conduct a campaign to counteract enemy propaganda and keep the hearts and minds of their men. Failure to do this will mean the rapid collapse of even the best-trained and best-equipped armies.

Tactical nuclear weapons may be of quite a small size. Usually the smallest weapon to be considered is roughly equivalent to 500 tons of T.N.T. This device has a relatively short radius of damage (less than 1,000 metres) and it can be fired from a large-calibre mobile gun, to a range of about 20 kilometres. Rockets will carry much larger nuclear warheads to any place on earth.

Aircraft may also deliver tactical nuclear weapons. Until recent years this has dictated the need for lengthy, carefully constructed airfields within striking distance of the battle area. These are difficult to defend and impossible to camouflage. At the start of hostilities they are a target of primary importance and their destruction will give the enemy air superiority. The Israelis demonstrated this in 1967. Now that Short Take Off and Landing (S.T.O.L.) and Vertical Take Off and Landing (V.T.O.L.) aircraft are being developed it should be possible to position them in concealed hides along the forward formations. It is very costly in fuel for the present designs to take off vertically, and, although it can be done, a reduced weapon load may have to be accepted. If a short runway, in the region of a few hundred yards, can be constructed the aircraft is much more efficient. Thus we have a new system of weapon delivery that can respond quickly to a request for a strike or reconnaissance mission.

The response times of nuclear guns and rockets are somewhat slow by comparison with conventional weapons. This is largely due to the care that must be taken in the preparation for firing and in giving warning to one's own side: troop safety is decidedly a limiting factor in the use of nuclear weapons. There may be times when, tragically, our own men will have to be sacrificed if it is impossible to withdraw them from the danger area.

Nuclear delivery systems will themselves be a primary target and this means that a policy of 'shoot and scoot' must be adopted after each firing. The size of some current equipment would make this difficult to achieve in a damaged war zone, unless whole regiments of engineers were available, which is not very likely.

Commanders require accurate and up-to-date information to enable their nuclear strikes to be planned. To provide this there is need for a coordinated, efficient system of intelligence gathering throughout the entire battle area and fairly sophisticated systems are now in general use.

Computers will help to collate, process and, after interpretation, disseminate the information. The whole emphasis is on speed and today one of the slowest parts of the system is man himself. Commanders at all levels must think clearly and quickly and be capable of making decisions speedily. The old maxim of leadership 'calm in a crisis and decisive in action' is very true on the nuclear battlefield. Let no one think that an army requires to be led by 'Boffins'. The traditions and training of Sandhurst and West Point are as important as ever. It cannot be doubted that the Duke of Wellington would have proved equally successful today, for he was a man with an eye for opportunity. Modern war requires modern Wellingtons, with a background knowledge of the technical weapons at their disposal.

To protect themselves, formations must be mobile. They should be able to disperse at a moment's notice to avoid annihilation or concentrate quickly to strike an enemy caught off balance. Armour will protect them, but they should still be able to dig into mother earth, when the situation demands. Nuclear radiation is rapidly absorbed by soil! As always commanders must do everything possible to deceive the enemy and conceal their true intentions.

So we have established that to be effective in nuclear combat troops must be armoured, mobile and capable of rapid concentration and dispersion. To make this possible, good communications are vital. Since 1945 military signals have advanced considerably. Today very high frequency and ultra high frequency systems, telex and television are all part of the expanding communication network. The ability to intercept or jam enemy radio transmissions is a 'weapon' in itself and the art of electronic warfare is rapidly becoming most important.

Radio signals are used in the guidance of missiles and direction-finding, as well as for pure communication. They can also play an important part in deception plans. Without good communications, the command and control required in modern war cannot be achieved.

Headquarters will become casualties and thus command arrangements must be flexible and duplicated. Units should have directives that will enable them to carry on for some time after their communications with higher authority have been destroyed. It is wise to set up alternative headquarters to take over in an emergency.

Standing still to fight in the nuclear battle will be to invite destruction. Units should adopt the maxim that the best method of defence is to attack. There should be no clear distinction between different phases of war and troops must be endowed with the offensive spirit. They should realize that safety from enemy nuclear weapons is found in close contact with the opposition or in rear of the enemy.

Whilst the bulk of our vehicles move on the surface of the earth as opposed to being airborne, mobility is an elusive characteristic. Swamps, lava belts, mountains and water obstacles will all tend to canalize movement to bridge or fords. Detailed reconnaissance and greatly increased engineer support is needed. Engineers building bridges will be speedily annihilated and thus there is a requirement for sappers equipped with amphibious bridging and rafts. These may be rapidly deployed, used and hidden again. There are never enough sappers! So as many vehicles as possible should be able to swim or wade.

Under the cover of darkness the armies must be resupplied and carry out maintenance of equipment and vehicles. Bridges and roads must be repaired at night and obstacles breached or strengthened. During these hours infantry will seek to infiltrate and this will call for the defence of scattered logistic units in rear areas. Long winter nights may even be an advantage!

Training at night has been neglected in Western armies, but not by the communist bloc. Their forces are well equipped, highly mobile and have an amphibious capability. They spend much of their time training at night and practise both chemical and nuclear warfare.

In the West there is an assumption that the communist forces will attack first. This means that in Europe there is a system of pre-planned defence. The details of this could not be published, but it is to be hoped that we do not depend on a 'thin red line'. After all, 'defence in depth' is an old tactical principle. If a natural obstacle can be improved by the addition of mines, booby traps and local inundation, an obstacle 'belt' may be built. This will not stop the invader, but if properly constructed it should slow down his advance long enough for the defending forces to deploy to battle positions. The 'belt' should be sited back from any frontier and highly mobile covering troops should be deployed along the border. As the enemy attacks, these troops should fall back ahead of him. Their aim must be to report the aggressors' advance and try to determine his strength and the direction of his thrusts. Where possible covering forces should impose delay, but they must not get themselves inextricably engaged. Armoured forces with helicopters in support are most suitable for this task. Finally, on reaching the obstacle belt they pass through pre-planned 'safe lanes' which are then closed behind them.

While the attacker breaches the obstacle he may present a nuclear target and persistent gas may also be used to thicken the belt – always assuming that the Governments concerned have given their approval to the use of such weapons. There are grave doubts whether in practice the democracies of the West would make up their minds in time. Indeed it would be to the advantage of the communist forces, if we did not use weapons of mass destruction. With their overwhelming man-power, they easily outnumber us and have an excellent chance of success in conventional war.

Whilst the attackers are still on the obstacle a thin line of troops equipped with surveillance devices (i.e. infra-red and radar) should observe their progress and pass information to their commander. He, in turn, should prepare to counter-attack with nuclear fire or, by using a light mobile screen, attempt to channel the penetrating enemy thrusts into a selected killing zone. This area should be like a *cul-de-sac* and once in it the enemy finds his progress barred by natural obstacles and minefields. Whilst he is halted in the killing zone he may present a nuclear target and a weapon should be already aimed at this point in readiness. The pre-planning of nuclear strikes will save valuable time. After the explosion of the device, the defender should counter-attack quickly to mop up the survivors. If no nuclear weapon is used the defender can only attempt to destroy his adversary by conventional means.

In a fluid battle of this nature, formations must be extremely flexible and units should not aim to hold ground or man defences in the obstacle zone. The defenders should try to destroy or break up the enemy's forces by continuous attacks, thus weakening him before he reaches the main defence. Units will have to accept much greater frontages and areas than ever before and must be prepared to be temporarily cut off for days at a time. They should accept penetration of the defences, but formations must have strong armoured mobile counter-attack forces ready to smash major thrusts.

In nuclear war the ebb and flow of battle will make it difficult to differentiate between attack and defence, defeat and victory, but when a deliberate attack is mounted it should be a concentrated and carefully coordinated blow delivered from positions in depth. Every effort should be made to gain surprise and strike the enemy where and when he least expects it. Deployment drills and battle procedures must be aimed at getting formations to concentrate, move, deploy and attack without delay. The technique of deploying into an assault formation from the line of march is already well known in many armies. After the attack, units must move on quickly and not present an opportunity for the enemy to use his nuclear weapons on them, should they constitute a worthwhile target.

The administration and supply during such a war is a

nightmare. Depots, hospitals and workshops must be hidden, but must also be able to move quickly and therefore need their own cross-country vehicles. Logistic units may well be nuclear targets and lacking manpower or earth-moving plant they are not easy to conceal underground quickly.

Movement on the battlefield will be hindered by the effects of nuclear 'blow-down'. This will take the form of fallen trees, destroyed buildings, broken bridges and roads jammed with burnt-out vehicles. Amongst this jumble there will struggle survivors, casualties, refugees, war correspondents and perhaps even the odd M.P.!

These theories are preached throughout the military academies of the world, but it is a very different matter to put them into practice. Mobility is by no means easy to achieve and amphibians, helicopters and hovercraft grow more expensive daily. We have learnt to depend on good communications and air superiority. One wonders how a sophisticated army may cope with a simpler, more self-sufficient force, operating in difficult terrain with few, if any, heavy weapons. Vietnam has given us an insight into the problems of the technological mind grappling with an assault by mass manpower. In countryside where tanks and A.P.C.s are useless and all the technology of the twentieth century cannot stop an army of millions, it may be found that the only defence is nuclear, biological or chemical.

For I dipt into the future far as human eye could see
Saw the vision of the World and all the wonder that would be,
Saw the heavens fill with commerce, argosies of magic sails,
Pilots of the purple twilight, dropping down with costly bales;
Heard the heavens fill with shouting, and there rained a ghastly dew
From the nations' airy navies grappling in the central blue.

Alfred Tennyson in his poem 'Locksley Hall', written over 100 years ago, may have foreseen the future.

# Index

## More about Penguins and Pelicans

*Penguinews*, which appears every month, contains details of all the new books issued by Penguins as they are published. From time to time it is supplemented by *Penguins in Print*, which is a complete list of all available books published by Penguins. (There are well over four thousand of these.)

A specimen copy of *Penguinews* will be sent to you free on request. For a year's issues (including the complete lists) please send 30p if you live in the United Kingdom, or 60p if you live elsewhere. Just write to Dept EP, Penguin Books Ltd, Harmondsworth, Middlesex, enclosing a cheque or postal order, and your name will be added to the mailing list.

Note: *Penguinews* and *Penguins in Print* are not available in the U.S.A. or Canada